Exploring Creation

with

Astronomy

by Jeannie K. Fulbright

Exploring Creation With Astronomy

Published by
Apologia Educational Ministries, Inc.
1106 Meridian Plaza, Suite 220
Anderson, IN 46016
www.apologia.com

Manufactured in the United States of America
Third Printing 2006

ISBN: 1-932012-48-6

Printed by The C.J. Krehbiel Company, Cincinnati, OH

Cover photos courtesy NASA/ JPL/ Caltech *Cover design by Kim Williams*

All Biblical quotations are from the New American Standard Bible (NASB)

Need Help?

Apologia Educational Ministries, Inc. Curriculum Support

If you have any questions while using Apologia curriculum,
feel free to contact us in any of the following ways:

By Mail: Dr. Jay L. Wile
Apologia Educational Ministries, Inc.
1106 Meridian Plaza, Suite 220
Anderson, IN 46016

By E-MAIL: help@apologia.com

On The Web: http://www.apologia.com

By FAX: (765) 608 - 3290

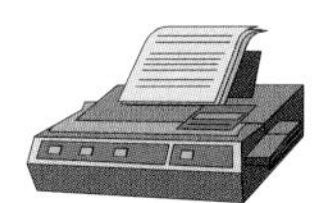

By Phone: (765) 608 - 3280

Illustrations from the MasterClips collection and the Microsoft Clip Art Gallery

Introduction

Congratulations on your choice of *Exploring Creation with Astronomy* as your science course this year. You will find this to be an easy-to-use science program for elementary students. The book is written directly to the student, making it appealing to kids. Presenting science concepts in a conversational, engaging style makes science enchanting and memorable for your students. It also fosters a love for learning. This course is written for children between six and twelve years old. When the course is complete, this book will serve as an excellent reference for your family's future questions and studies in astronomy. As a result, this book is a life-long investment!

The Immersion Approach
"Is it Okay to Spend a Year on Astronomy?"

Many educators promote the spiral approach to education. In this approach, a student is exposed over and over again to minute amounts of a variety of science topics. The theory goes that we just want to 'expose' the student to science at this age, each year giving a bit more information than was given the year before. This method has proven unsuccessful in public and private schools. It assumes the young child is unable to understand profound scientific truths. If a child is presented scant and insufficient science, she will fail to develop a love for the subject, because it seems rather uninteresting. If the learning is skimpy, the subject seems monotonous. The student is simply scratching the surface of the amazing and fascinating information available in science. Sadly, students taught in this way are led to believe they "know all about" that subject, when in reality the subject is so much richer than they were allowed to know or explore. That is why we recommend that kids, even young kids, are given an in-depth exposure into each science topic. You, the educator, have the opportunity to abandon methods that don't work so that your students can learn in ways that are more effective. The immersion approach is the way everyone, even young kids, learn best. That is why we major in one field in college and take many classes in that field alone. If you immerse your student in one field of science for an entire year, he will develop a love for both that subject and a love for learning in general.

On the other hand, if a student rushed through several fields of science in one year, she will feel insecure about the information. In fact, she probably doesn't really know much science, and that's why she feels insecure. Imagine the benefit to your student when she is able to authentically converse with a scientist at the planetarium, intelligently discussing the dynamics and idiosyncrasies that are seen in the universe. This will delight both your student and adults with conversation that is actually interesting and intelligent. A child taught with the immersion approach learns to love knowledge and develop confidence.

Additionally, a child that is focused on one subject throughout an entire year is being challenged mentally in ways that will develop his or her ability to think critically and retain complex information. This will actually benefit the student and give him an advantage on achievement tests. He will be able to make more intelligent inferences about the right answer on a science question, as God has created an orderly world that works very similarly throughout all subjects of science. A child who has not been given the deeper, more profound information will not understand how the scientific world operates, and he will not be able to reason through the problem to get the correct answer.

Lesson Increments

The lessons in this book are in-depth and contain quite a bit of scientific information. Each lesson should be broken up into manageable time slots depending on a child's age and attention span. This will vary from family to family. There are 14 lessons in this text. Most lessons can be divided into two week segments; reading the text and doing the notebook assignments the first week, and working on the activities and the projects the second week. If you do science two or three days per week, you can read four to six pages a day to finish a lesson and begin the activities and projects. This will give you 28 weeks for the entire book, allowing for leniency with vacations and family emergencies or lessons and experiments that take a little longer than expected. Lessons 2 and 13 will probably take longer than two weeks, for example, as they contain more information than the other lessons.

There are no workbooks or tests that go with this program. They are unnecessary at this age. Instead, the very effective method of narrating and keeping notebooks is used. This is a superior approach that facilitates retention and provides documentation of your student's education.

Narrations

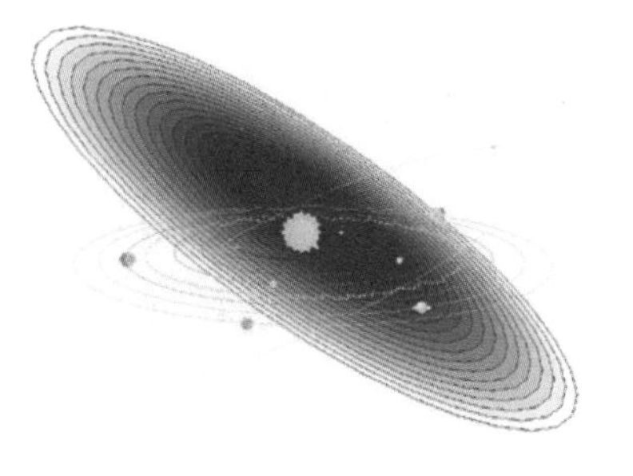

Older elementary students can do the entire book and most of the experiments on their own, while younger students will enjoy an older sibling or parent reading it to them. Each lesson begins with reading the text. Throughout the reading, the student will be asked to retell or narrate the information she just studied. This helps the student to assimilate the information in her mind. The act of verbalizing it in her own words propels her forward in her ability to effectively and clearly communicate the information to others. It also serves to lock the information into her mind.

Please don't skip the narrations. Though they may seem to take up valuable time, they are vital to your students' intellectual development. The more narrating the student does, the better at it he will become. The better the student becomes at narrating, the better he will be at writing, researching, and clearly communicating his beliefs. Some teachers encourage their students to take notes as they listen to reading. You may or may not want to try this.

The narration prompts are usually found under the title "What Do You Remember." At the end of the day's reading, your student should write down or dictate to you what she remembers from the lesson. This written narration will be put in her notebook. At that time, you will have your student make an illustration to accompany her written narrative. If the student habitually restates her learning, adding these accounts to her astronomy notebook, she will accrue a volume of her own work that can be referred to repeatedly throughout the year.

Notebooks

In addition to the written narrative, a notebook activity concludes each lesson to promote further experience with the material. These activities are often entertaining and sometimes challenging to the students. Each notebook activity requires the student to utilize the material he learned in a creative way that will further enhance his retention. The notebook activities generally occur at the end of a lesson. It is recommended that you also include a narrative of any related field trips, projects, or activities.

Encourage your student to treasure his notebook, seeking to do his best work. His notebook is an important tool which will provide you and your student with a record of his course work. It will also serve as a foundation for future studies when the student is ready to take learning to the next level. Your student's notebook will spark lifelong memories of his homeschooling experience. Require diligence regarding his notebook.

Latin Root Words

There are a great many Latin words in science. Latin is the foundation of many languages, and knowing Latin roots aids a student extensively in understanding our language. This book will often define the Latin terms, breaking the word down into its root forms. It would benefit your student to take notes when she reads and copy all the Latin roots she learns into a separate Latin notebook. This notebook will serve the student well throughout her academic years, as it will provide the groundwork for later language study.

Course Website

When you look at the night sky, what you will see depends on where you are on the earth. As a result, it is difficult to write an astronomy book that applies to everyone. In addition, many astronomical events are rare, and it would be hard to continually update a book to give you the future dates for these interesting events. To get the most out of this course, you should regularly visit the course website. It will give you links that tell you how to find things in the night sky for your area and the time that you are covering astronomy. It will also give you future dates for upcoming astronomical events. Finally, it will also give you links to interesting multimedia that relates to the book. To go to the course website, simply type the following address into your web browser:

http://www.apologia.com/bookextras

You will see a box on the page. Type the following password into the box:

Godcreateditall

Make sure you capitalize the first letter, and make sure there are no spaces in the password. When you hit "enter," you will be taken to the course website.

Projects and Experiments

Each lesson contains hands-on projects, activities, and experiments, bringing the concepts to your student's level of comprehension. Though they may seem time consuming, these projects and experiments are important to the subject and will benefit your student immensely. They play an important role in the younger student's learning. Please try very hard not to skip them, as they will increase your student's understanding of the subject.

The projects and experiments in this book use common, household items. As a result, they are fairly inexpensive, but you will have to hunt down everything that you need. To aid you in this, the next two pages list the materials that you need for the experiments, activities, and projects in every lesson. If you would rather spend some money for the sake of convenience, you can purchase a kit that goes with this course. It is sold through a company called "Creation Sensation," and you can contact them at 501-776-3147. Alternately, you can visit their website at http://www.creationsensation.com/.

Items Needed To Complete Each Lesson

Every child will need his own notebook, blank paper, lined paper, and colored pencils.

Lesson 1

- Balloons of many sizes and colors (Very small balloons, like water bomb balloons, and very large balloons, the bigger the better)
- Scissors
- Thumbtacks
- Thread, ribbon, or string
- Markers or paint
- Measuring tape (If you do not have a measuring tape, you can use string cut to the lengths listed in the project.)
- Construction paper

Lesson 2

- A magnifying glass
- A refrigerated chocolate bar (Any candy bar that has a lot of chocolate in it will do.)
- A flashlight
- A globe (or round ball)
- A Styrofoam® ball (or any small ball) on a string
- A box
- Scissors
- White paper
- A pin or needle
- Tape
- Aluminum foil

Lesson 3

- A large bowl
- White flour
- Several pebbles of different size
- A large bowl
- A marble or pebble
- A pencil
- ¼ cup of salt
- 1 teaspoon cooking oil
- ¼ cup of water

Lesson 4

- ¼ stick of butter or margarine
- Some flour
- A large plate
- A small bowl (like a custard dish) or a small cup
- A saucepan
- A stove
- A shoe box or other square box
- A piece of cloth, such as a dish cloth or strong paper towel, that will cover the box and drape down its side.
- A large rubber band and perhaps some tape
- Newspaper
- Plaster of Paris or flour and water
- Bamboo skewer (or a long, skinny stick with a point)
- Markers, crayons, or colored pencils (Markers work best.)
- Ruler or measuring tape

Lesson 5

- A baseball
- A tennis ball
- Lamp
- Globe
- A cork
- A permanent marker
- A lid from a yogurt or sour cream container (Try to find one that has a high lip.)
- A sewing needle
- A magnet

Lesson 6

- Lamp
- Lightly-colored ball
- Compact disc (CD)
- Two magnifying glasses (One should be stronger than the other. Reading glasses will work also.)
- Cardboard tube (Mailing tubes, paper towel rolls, and wrapping paper rolls all work well.)
- Strong tape (Masking tape and duct tape work best.)
- Scissors (For thick cardboard tubes, you might need a knife.)
- Tape measure
- Paper with writing or an image on it

Lesson 7

- Construction paper (Some of it should be green.)
- Sticks
- Long, thin strips of cardboard
- Plastic cling wrap or wax paper
- A small plastic bottle, such as a vitamin or aspirin bottle
- A large baking pan (not a cookie sheet) or aluminum foil
- Vinegar (White vinegar works best because it won't interfere with the color.)
- Rocks (optional)
- Baking soda
- Red food coloring
- Dish washing liquid
- 1½ cup of white flour
- ¼ cup of salt
- 1 tsp. cooking oil
- ¼ cup of water

Lesson 8

- A ball that is about the size of a soccer ball
- Three tiny round sprinkles or large grains of salt or sand
- Several sugar sprinkles that are smaller than the round sprinkles above
- Two "red hots" (red cinnamon pieces for cookie decorating) or two small, whole allspice
- A pecan or large marble
- An acorn or large marble (a bit smaller than the one above)
- Two pinto beans or lima beans, one smaller than the other

Lesson 9

- Two plastic soda bottles
- Electrical or duct tape
- 1-inch washer
- Water

Lesson 10

- Alka-Seltzer tablet®
- Eye protection
- Film canister (inside enclosure)
- Water
- Tape
- Paper
- Scissors

Lesson 11

- A ball (no larger than a tennis ball)
- A glass jar
- A match
- Ice
- A Ziploc® bag large enough to cover the opening of the jar
- Hot water

Lesson 12

- 2 tablespoon powdered sugar
- ½ cup whipping cream (Whole milk or half-and-half will work.)
- ¼ teaspoon vanilla
- 6 tablespoons rock salt
- 1 pint-size Ziploc® plastic bag
- 1 gallon-size Ziploc® plastic bag
- Several ice cubes

Lesson 13

- Cardboard
- Clear plastic wrap
- Tape
- Scissors
- A shoe box
- A marker
- A bamboo skewer or ice pick
- A flashlight
- Scissors
- A dark room

Lesson 14

Creative materials for building a space station model, such as:

- Wires
- Paper towel rolls
- Empty soda bottles
- Craft sticks
- Straws
- Lids

Exploring Creation With Astronomy

Table of Contents

Photograph and Illustration Credits

Photos and Illustrations courtesy NASA/JPL/Caltech: Cover, 1, 2 (picture of stars and comet Halley), 6, 7 (Saturn V), 8 (both), 11, 15, 16 (solar flare photo with earth inset), 17, 24, 29, 30 (Mercury), 30 (Mercury part of sun/Mercury illustration), 32 (Mercury part of sun/Mercury illustration), 32 (Mercury and earth), 33 (all three – asteroid crash is a montage), 34, 35 (both), 38 (both), 40 (bottom image), 42, 44 (both), 45 (volcano), 46 (planets in montage), 51, 52, 56, 57 (montage), 58, 59, 60, 61 (sled), 62 (magnetosphere), 65, 67 (earth and moon), 69 (moon during eclipse), 70, 71, 72, 77, 78, 79 (all), 80 (both), 81 (happy face crater and moons), 82, 83 (both), 84, 87 (both), 89, 90 (Schwassmann-Wachmann), 91 (coma), 92, 93, 94 (all), 95, 96 (both), 98, 99, 103, 104, 105 (red spot), 106-109 (all), 112 (red spot), 113, 114, 115 (both), 117 (both), 119 (both), 121-127, 131, 132 (bottom), 133-135 (all), 137 (top illustration is a montage), 141,142 (Photo by Andy Steer), 145 (Drawing by A. Hobart), 146, 150 (all), 155 (bottom), 159-169 (all), 172

Photos and illustrations from www.clipart.com: 2 (drawing of telescope and stars), 3, 4 (both), 5 , 7 (illustration of orbits), 13, 14, 16 (drawing of earth rotating), 20, 21 (sun illustration), 23, 28 (sun), 30 (sun part of sun/Mercury illustration), 32 (sun part of sun/Mercury illustration), 46 (sun in montage), 48, 53 (montage), 54, 63, 64, 66, 67 (hand w/CD), 68 (montage), 69 (sun, earth, and moon montage), 73 (both), 81 (Arizona crater), 90 (Hale-Bopp), 92 (dust trail), 97 (montage), 105 (lightning), 129, 143 (bottom), 144 (montage), 153, 154, 156, 170

Photos and illustrations by Jeannie K. Fulbright: 10, 18, 19, 21 (waves), 26, 27, 28 (box), 50, 76 (both), 86, 88, 112 (bottles), 120 (both), 139, 157, 158

Photos and illustrations by Dr. Jay L. Wile: 12, 41, 91 (orbits), 130, 132 (orbits), 136

Digital artwork by Dr. David Heatley: 31, 40 (top illustration), 116

Photo by Mark Whitney: 43

Photo of Venera 13 (page 45) is from the Russian publication Space Science, vol 21 #2, pp. 176-99. Used with permission.

Photos and Illustrations courtesy NOAA: 55, 62 (Northern Lights)

Illustration by Megan Whitaker: 62 (earth)

Photos by Bill and Sally Fletcher: 143 (dippers), 148, 152 (all)

These photos were taken using a special technique developed by the photographers. This technique "expands - in their true color - the bright stars that make up the constellation's shape. This is done while leaving the background stars and deep space objects unaffected. It is a natural photographic process accomplished during the exposure of the film." (http://www.scienceandart.com/0galleryconst.htm, retrieved 5/16/2004)

Illustration from the MasterClips collection: 155 (top)

Lesson 1
What is Astronomy?

What Is Astronomy?

"The heavens are telling of the glory of God; and their expanse is declaring the work of His hands."

Psalm 19:1

The Bible calls everything up in space "the heavens." Everything up in the heavens is God's; it all belongs to Him, and it was all made by Him and for Him. The more we learn about all that is up in outer space, the more amazed we become at how perfectly God created the **universe** (you' nuh vurs), which is the earth, planets, sun, stars, and everything in space. The universe really does show us how sensational God is, and when you are finished with this book, you will be even more amazed than you are right now.

The study of the universe is called astronomy (uh strahn' uh me). The word "***aster***" means "***star***," while "***onomy***" means "***knowledge of***." The word astronomy, then, means "knowledge of the stars." Many years ago, the only word used for every object in outer space was aster, or star. In other words, every light in the night sky was called a star. We still use the word astronomy to talk about the study of everything in space, even though the way we use it means more than just studying the stars. An astronomer is someone whose job is to study the stars, the planets, and everything out in space. You are going to be an amateur (which means beginner) astronomer this year, because you will be studying the universe as you take this course.

Have you ever been out in the country at night? You can see more stars in the sky out there because you are far from the lights of the city. Even people who live many miles away from a big city still cannot see all the stars in the night sky because of the city lights in the distance. On a clear night in the country, you can see many thousands of stars up in the sky. It's truly a miraculous sight. When the Bible was written, people could see more of the stars than we can now because there were no bright lights to drown out the lights in the sky. When people looked in the sky and saw how enormous the heavens were and how bright the stars covering every inch of it were, they knew that the Creator of the world was marvelous and mighty.

This is a picture of comet Halley (the bright spot near the center of the picture) with the stars of the Milky Way in the background. You will learn about comets and the Milky Way later on in the course.

Why Did God Create the Universe?

The first chapter of the book of Genesis tells us that the heavens and the earth were created in six days. God took a great deal of care with the earth before He created other things. In the first three days, God created the oceans, mountains, plants, and trees. Then, on the fourth day (before He made any animal on earth), He made the stars and planets in the sky. Genesis 1:14-15 tells us, "Then God said, 'Let there be lights in the expanse of the heavens to separate the day from the night, and let them be for signs and for seasons and for days and years; and let them be for lights in the expanse of the heavens to give light on the earth.' And it was so."

There are many reasons God made the universe. God made the stars and the moon to give us light at night and to give us a calendar and signs. Scientists have also learned that the planets, stars, and many other things in space help to keep life going on the earth. All of these things glorify God because only God could have made these things the perfect way they are made.

Calendar

God made the planets, moons, and stars in the sky so that we could tell time, know the timing of the seasons, and count the days and years. Many years ago, before people had calendars and clocks, they told the time of day by the position of their shadow on the ground. They also knew when a month had passed by the shape of the moon in the sky. There is a monument in England called **Stonehenge**. Many believe that ancient people used it to tell when spring had arrived. They judged the season by the position of the sun in relation to the large stones that make up the monument. Knowing what day spring began helped them to time the planting and harvesting of crops.

This is a picture of Stonehenge, an ancient monument in southern England. We are not sure what its purpose was, but many historians think that it was an ancient structure used to track the seasons.

The patterns of stars in the sky are called **constellations** (kahn' stuh lay' shuns). Ancient people also knew which constellations would be in the sky in each season of the year -- winter, spring, summer, or fall. They also used the constellations to mark what year it was and how many years had passed since an event. Many years ago, then, before we had calendars in our homes, the stars were the only calendar.

Years ago sailors did not have compasses as they do now, so the stars helped them to know which direction to sail. Now sailors use compasses. A compass is a device that has a needle that always points north. You always know what direction you are going if you have a compass, but you can also know which direction you are going if you know where the constellations are.

God's plan for the lights in the sky does not only include mankind. Scientists have learned that some birds know to fly south for the winter by the constellations. The birds must fly, or migrate, south for the winter. Otherwise, they would not have enough to eat, and sometimes they would get too cold. God made a very special way for birds to know when and how to fly south. He created within them a special gift we call **instinct** (in' stinkt). One instinct that God has given birds tells them to look at the constellations to know when to migrate south for the winter and when to migrate back north for the summer. It also tells them how to use the constellations to know *which direction* they must fly. This is why birds often fly at night when they migrate. When birds are not migrating, they usually sleep at night.

God's Signs

God placed a star over the city of Bethlehem when Jesus was born. This was a sign that the Savior had come. The sign was so powerful that when wise men from a distant land saw the star, they traveled to Bethlehem to see the baby Jesus. Some Christians believe that the constellations also once served to tell the story of our Savior. They believe that God placed constellations in the sky that originally told the story of Jesus, who would be born of a virgin, would die, and then would rise again. You see, in ancient times, not very many people were able to read. It is possible that they used the constellations as "pictures" which would remind them of what God was going to do one day by sending Jesus to die for all our sins. You will learn more about this interesting idea later on in the course. Even if this idea is true, we do not need the constellations to tell the story of Jesus anymore. After all, we now have God's Word in the Bible.

Stars and Planets

Not everything you see in the sky that looks like a star is actually a star. The two brightest objects in the night sky (besides the moon), for example, are planets. Several other planets can be seen in the night sky as well, and to the untrained eye, they all look like stars. However, a trained eye can tell the difference. Stars appear to twinkle in the sky, but planets do not twinkle. That is one way to tell if you are seeing a star or a planet. There are also stars that appear to move rapidly across the sky and then disappear. We call them "shooting stars." They are not stars at all. They are meteors (mee' tee orz). You will learn about all of these things later on in this course.

The planets help to fasten the earth in place. They keep the earth from moving too far away from the sun or too close to it. In other words, the planets make our world steady. The Bible told us thousands of years ago that God "established the earth upon its foundations, so that it will never totter forever and ever" (Psalm 104:5). Have you ever seen something on the playground called a teeter-totter? It is sometimes called a seesaw. To "totter" means to "sway and wobble." The earth would definitely totter if the planets were not exactly where they are. You see, the sun pulls on the earth with a force called **gravity**. This force holds the earth in its orbit around the sun. However, the earth's orbit would totter if it were not for some of the other planets. They also pull on the earth with gravity, and this "smooths out" the motion of the earth as it travels around the sun. Yes, indeed, the earth would wobble and shake in its motion around the sun if God had not put the other planets where they are. How wonderful that God made the planets up in the sky to help the earth stay right in the perfect place where God put it. The earth is a very important planet to God, and He made sure that nothing Satan could bring about would destroy it.

We can see now that God has a lot of very special reasons for making the stars and planets. I'm sure there are many more that we do not even know, but we can be sure that everything was done just perfectly according His plan. What a wise and wonderful God He is!

Solar System

Our solar system is the sun and all the planets that travel around the sun. God placed all the planets in their positions and planned it so that they are all held in place by gravity. Gravity is an invisible force that pulls objects towards each other. When we drop something, it doesn't really fall; it gets pulled down to the earth by gravity. Instead of saying, "it fell," it would be more scientifically correct to say, "It was pulled to the earth." All the planets and their moons have gravity. Heavier planets have more gravity than lighter planets. The sun, the heaviest thing in our whole solar system, has the most gravity of all. God placed the planets and sent them to circle around the sun at the perfect distance. If the planet Mercury (mur' kyur ree) were very much closer to the sun, or if the sun were very much heavier, Mercury would get pulled right into the sun. Instead, it stays exactly where God put it, because it has been placed at the right distance from the sun and because the sun is not too heavy. The pull that planets (and other objects) have on each other is called **gravitational pull**. The sun, earth, and all of the planets have gravitational pull. The earth has a gravitational pull on the moon, which keeps the moon right where it is. The moon has a gravitational pull on the earth, which makes the oceans bulge up toward it as it passes by. The sun has gravitational pull that keeps the planets in their places in the solar system.

There are nine planets that we know of in our solar system: Mercury, Venus (vee' nus), earth, Mars, Jupiter, Saturn, Uranus (yur' uh nuhs), Neptune (nep' toon) and Pluto. This is also the order in which they travel around the sun, as shown below:

This is a drawing that represents our solar system. Only part of the sun is shown, but each planet is shown along with where it is in relation to the sun. As you can see, Mercury is closest to the sun, while Pluto is farthest from the sun. Although the relative sizes of the planets are correct, the sun's size is not. It should be much larger. The distance between the planets is not correct, either.

The way that many people remember the planets and their order is by making a mnemonic (nih mahn' ik) phrase. In a mnemonic phrase, the first letter of each planet is made into a different word that makes a sentence. Look at this example:

Mercury	Venus	Earth	Mars	Jupiter	Saturn	Uranus	Neptune	Pluto
My	**Very**	**Early**	**Morning**	**Just**	**Started**	**Under**	**Nancy's**	**Pancakes**

Notice that the word underneath each planet begins with the first letter of the planet's name. The sentence that those words make helps you remember the order of the planets. That is a silly sentence, "My very early morning just started under Nancy's pancakes," but the first letter of each word in that sentence helps you remember the order of the planets. At the end of this lesson, you will make your own mnemonic phrase to help you remember the order of the planets.

Can you explain what you have learned so far in your own words?

Astronomers, Astronauts, and Satellites

There were many people in history that helped us understand astronomy better. Way back in the year 1510, a man named **Nicolas** (nik' oh lus) **Copernicus** (koh pur' nih kus) had the unusual and amazing idea that the earth revolved around the sun. At that time, everyone thought that all the stars and planets revolved around the earth.

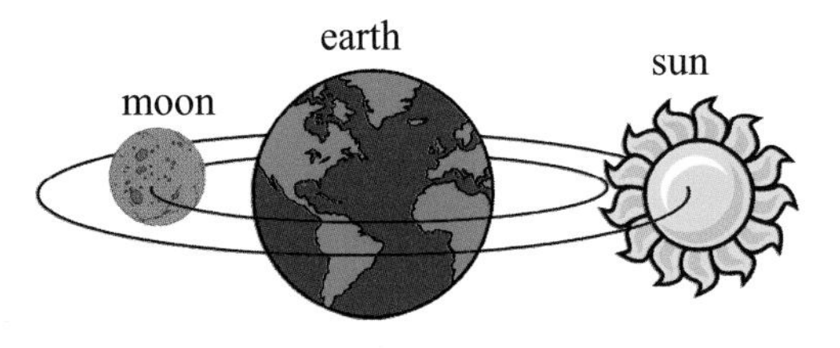

In Copernicus's time, everyone thought that the earth was the center of the solar system and that all of the planets and the sun revolved around the earth.

OR

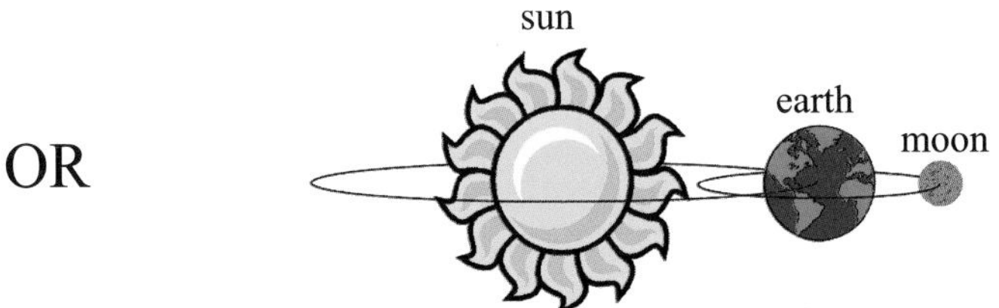

Copernicus thought that a more elegant arrangement of the solar system would be for the sun to be at the center and for the planets to revolve around the sun.

We now know that Copernicus was correct, even though most people during his time did not believe him.

Galileo (gal ih lay' oh) **Galilei** (gal ih lay') was an astronomer who believed Copernicus. He taught us to use telescopes to study the planets and stars, and many of the observations that he made helped scientists understand that Copernicus was right about the sun being at the center of our solar system. Galileo was able to learn a lot of things about our solar system through the wonderful telescopes he built.

Today, a lot of astronomers work for **NASA**. NASA is America's space program, and it stands for "**N**ational **A**eronautics and **S**pace **A**dministration." If you want to be an astronomer when you grow up, you might want to work for NASA. NASA is also the organization that sends people and spaceships to space. If you like to build things and make inventions, you could be a NASA engineer. NASA engineers build spaceships, telescopes, robots, and many other useful things for space exploration. The picture on the left shows a rocket being built by NASA engineers. See how tiny the engineers at the bottom of the picture are? That gives you an idea of how big the rocket is.

This is a picture of the Saturn V rocket being assembled by NASA engineers. The "V" in "Saturn V" is the Roman Number "5," which refers to the number of engines in the rocket's first stage.

If you became an astronaut, you will probably work for NASA. An astronaut is someone who is trained to travel in a spaceship into outer space. He or she uses a special spacesuit to explore outer space. Maybe one day you will be an astronaut and go to some of the places we will study in this course! Many astronomers, engineers, and scientists work for NASA.

This is a picture of an astronaut in a space suit.

Have you ever looked through a telescope? You can see a long way off when you do. You will see many pictures that come from telescopes as you study this course. There is an enormous telescope floating up in space that sends pictures back down here to earth. It is called the **Hubble Space Telescope**. Even though a telescope will make a planet look like it is much closer, most planets can be seen without a telescope if you know where to look. You can even see satellites. A satellite is an object up in space that travels in circles around another object. The moon is a satellite of the earth because it travels in a circle around the earth. So when you are looking up in the sky, you can say, "Oh look! I see a satellite!" as you point at the moon.

This is a picture of the Hubble Space Telescope in orbit around the earth.

An artificial satellite is something that is made by man and sent into space to float around the earth. "Artificial" means "not natural;" something that is made by human hands. Only God makes natural satellites. There are thousands of artificial satellites that people have sent up to travel around the earth. They do many jobs. Some can look closely at any part of the world and take pictures for others to see. Some can put a lot more channels on your TV. Some can look at planets and stars far away. Some watch the weather and take pictures so the weatherman can tell us if it is going to rain. Artificial satellites are very important. If you look up into the sky at night and see a small point of light (like a star) that is moving across the sky, you are probably looking at a satellite.

Can you tell someone what you have learned about astronomy so far? Remember to include what astronomy means and what gravity is. By telling someone what you learned, you "lock" it into your brain and will remember it longer.

What Do You Remember?

Why did God create the stars and planets? What are the names of the planets? What is the name of America's space program? What does NASA do? Do you remember the name of the astronomer who first said that the earth revolves around the sun? What about the name of the astronomer who learned how to study space with a telescope?

Your Notebook

You are going to make an astronomy notebook during this course. You will put all the things you make, draw, and write about astronomy in your notebook. You can also add things you learn from books, the Internet, and magazines. When you are finished with this study, you will have a complete notebook on astronomy that you can read whenever you want to remember what you have learned.

You will make many illustrations for your notebook. Illustrations are pictures you have drawn. Please take your time when you make your illustrations. Do a good job so that you can be proud of your work in the years to come. One day you will look at your pictures and be amazed at what a great job you did. This will only happen if you are careful with your work, and not in a hurry.

Assignments

Now it's time to design the front cover of your astronomy notebook. Remember that this is your very own notebook, so you may put whatever you wish on the front cover.

Make a mnemonic phrase to help you remember which order the planets come in. List the planets on a piece of paper like this:

Mercury	Venus	Earth	Mars	Jupiter	Saturn	Uranus	Neptune	Pluto

Write in the box below each planet a word that begins with the first letter of the planet listed. Try to make a sentence that you will remember. It will be easier to remember a sentence that makes sense. You can also make more boxes so that you can make more mnemonic phrases. Place the phrases in your notebook.

Project

Build a model solar system that hangs from your ceiling

You are now going to build a model of the solar system. Every model has its weakness. That means there is always something that makes it imperfect and unlike the real thing. That is the nature of models, including the one you are going to make. In this model, you will get a good feel for how big each planet is compared to the others and where they are placed in relation to the sun. Unfortunately, in the model you make, the planets will not be as round as the real planets are, and the distances between the planets will not be realistic. In addition, the size of the sun will not be correct at all. If you were to build this model so that the size of the sun was correct relative to the other planets, the sun would simply be too big to hang from the ceiling.

You will need:

- Balloons of many sizes and colors (You will need very small balloons, like water bomb balloons, and very large balloons, the bigger the better.)
- Scissors
- Thumbtacks
- Thread, ribbon, or string
- Markers or paint
- Measuring tape (If you do not have a measuring tape, you can use string cut to the lengths listed in the project.)
- Construction paper

Instructions

1. You will start by choosing several balloons, each of which will represent a planet. For each planet, try to choose a balloon with a color that is something like the planet's color. Mercury should be somewhat gray, for example, while earth should be blue. Use the drawing on page 6 to help you decide the color for each planet.
2. Once you have chosen a balloon for a planet, slowly blow up the balloon. As you blow it up, have your parent / teacher measure the distance around the balloon at its widest point. Tie the balloon closed when the distance around its middle is given by the table below:

Planet the Balloon Represents	Distance Around the Balloon	Planet the Balloon Represents	Distance Around the Balloon
Mercury	1 inch	Saturn	25 inches
Venus	2 $\frac{1}{2}$ inches	Uranus	10 $\frac{5}{8}$ inches
Earth	2 $\frac{5}{8}$ inches	Neptune	10 $\frac{1}{4}$ inches
Mars	1 $\frac{3}{8}$ inches	Pluto	$\frac{1}{2}$ inch
Jupiter	29 $\frac{1}{2}$ inches		

Use markers or paint to decorate the planet so it looks like what is in the drawing on page 6.

3. Make a circle out of construction paper that will fit around the balloon that represents Saturn.
4. Tape that circle to the balloon to represent Saturn's rings.
5. Blow up the largest balloon you have as big as you can possibly make it to represent the sun.
6. Tie a string, ribbon, or thread to each balloon, and use the thumbtacks to hang the balloons from the ceiling. Make sure you hang the balloons so that they are in the correct order. Use the drawing on page 6 or the picture above as a guide.

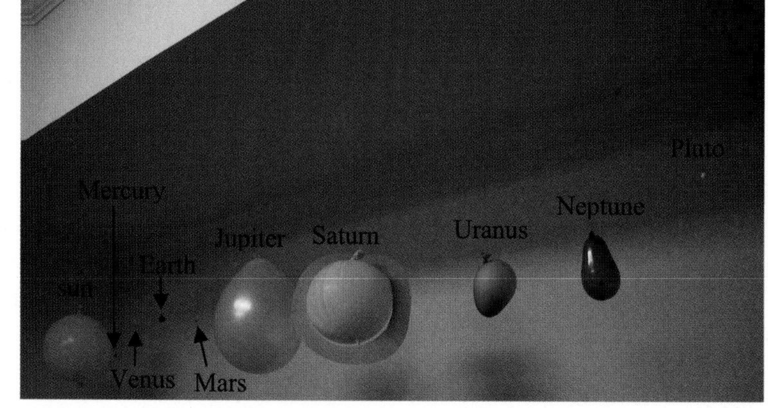

Lesson 2
The Sun

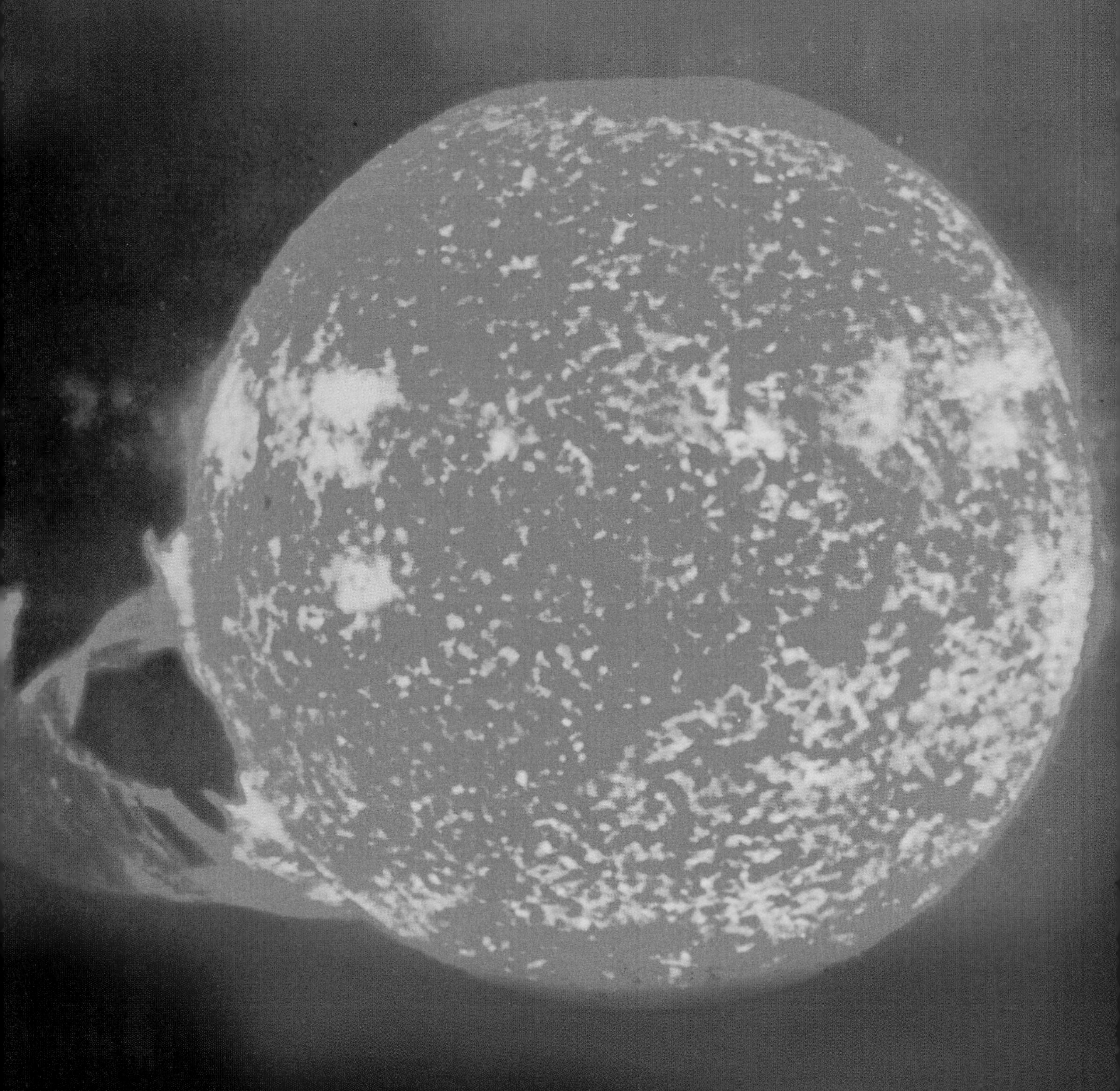

The Star of Stars

The sun seems so small up in the sky, but it is one of the biggest things God created. It is so enormous, so gigantic, that you have never seen anything so big in all of your life. It's bigger than the biggest thing on earth. What's the biggest thing you can think of that you have seen on the earth? Did you know that the sun is millions of times bigger than that? The sun makes everything on earth, and even most things in space, look like tiny little specks of dust or tiny little ants. The difference in size between the earth and the sun is very great. You can see the difference if you compare a small peppercorn to a basketball. If you don't have a basketball, a dinner plate will work also. Just remember that the sun is not flat like a plate.

If the sun were the size of the basketball in this picture, the earth would be the size of the peppercorn on its right.

Now look at these two objects next to each other. The peppercorn represents the earth, and the basketball or plate represents the sun. Amazing isn't it? One million earths would fit inside the sun! The whole entire earth is like a tiny speck compared to the sun. If the earth is so small compared to the sun, imagine a mountain on the earth compared to the sun. Imagine a person compared to the sun. Imagine an ant compared to the sun!

Why does the sun seem small in the sky if it so much bigger than the earth? Shouldn't it look big to us since we are tiny in comparison? Well, look out the window. Find something on the earth, a house or tree or mountain that is really far away, as far away as you can see. Then close one of your eyes and hold your finger up next to that far away object. Do that now.

Doesn't it look like that object is smaller than your finger? You know that the object is much larger than your finger, but your finger seems bigger is because it's closer. The closer an object is to us, the bigger it will seem. Even big objects look tiny when they are far away.

92,935,700

Can you say that number? It will help if you call the first comma "million" and the second comma "thousand." That is a giant number. Ninety-two million, nine hundred thirty-five thousand and seven hundred is how we say it. That's how many miles we are away from the sun. That is further than you can even imagine. Most people "round up" and just say ninety-three million miles.

How many miles did you travel last time you went on a trip? Did you go 100 miles? 700 miles? I bet you didn't go ninety-three million miles. It would take years and years to get to the sun if we tried to travel there. But we wouldn't ever do that, because it is just too hot!

The sun is alight with heat and swirling gases. God made the sun so hot that we can feel its heat millions of miles away. The hottest your oven can get is about 500 degrees. The sun is usually about 10,000 degrees on the outside and millions of degrees on the inside! If something were as hot as the sun and it touched the earth, it would burn a hole all the way through in an instant. God put the sun the perfect distance from the earth. If it were closer, the water in the oceans would evaporate. The trees and plants would all die, and we would all burn up. If the sun were further away, its warming energy would not reach us, and the oceans would freeze into big blocks of ice, and so would we.

Don't Stare!

The sun's light is so powerful that it isn't even safe to look at the sun! It can cause terrible damage to your eyes if you stare at the sun. Have you ever looked at a bright light bulb and then had to look away after a moment? Well, the sun is about a million times brighter than a light bulb. This is why you can injure your eyes if you look directly at the sun. The sun is so bright that it wouldn't even be safe to look right at the sun if you were standing on Pluto, the furthest planet from the sun! To study the sun, scientists look at it with the help of special tools.

Did you know that you can take a magnifying glass out into the sun and burn little holes in leaves? This works especially well during the summer when the sun is shining its light more directly upon us. When you focus the sunlight coming through the curved lens of the magnifying glass for a short period of time, it will burn a hole in the leaf. This happens because the rays of light get bent by the curved lens and concentrated to a small spot, focusing all of the heat and energy there. That's a lot of heat and energy to be going to one spot. That's why it burns a hole in the leaf. Well guess what! You have a lens just like a magnifying glass in your eye! If you look at the sun, your eye-lens will concentrate the sun's light and focus it to a very small spot on the back of your retina (the back wall of your eye.) This can cause permanent eye damage or blindness. Also, there are no pain sensors in

your retina, so you wouldn't even know it's happening! Please do not ever look directly at the sun, but you are welcome (with your parent's permission) to take a magnifying glass to a leaf!

Can you explain what you have learned so far in your own words?

Revolve and Rotate

When one object travels in a circle around another object, it is orbiting that object. In other words, it is revolving around that object. We use the word "orbit" or "revolve" to mean the same thing. One object is in the center, while the other object moves in a circle around it.

The earth *revolves* around the sun

Have you ever heard it said that someone's life or world revolves around something or someone? One person might say, "Her life revolves around ballet." Another person might say, "That mother's world revolves around her baby." That is just an expression. It means that something is very important to someone. What does your life revolve around? It is best for all our lives to revolve around Jesus, for He should be the Ruler of our lives!

Many years ago, astronomers and everyone else believed that the sun revolved around the earth. Every morning they would see the sun coming up in the east and going down in the west. It just seemed logical that it was circling around the earth.

But now we know that the sun is in the center of the solar system, and the earth revolves, or orbits, in a circle around the sun. Remember that a satellite is an object in space that revolves around another object. The earth is a satellite of the sun, just as the moon is a satellite of the earth! The sun has many satellites, including the nine planets and thousands of asteroids, comets, and meteoroids. The earth has only one natural satellite (the moon), but there are thousands of artificial satellites orbiting the earth.

Take a Walk Around the Sun

Here is an activity to try, but you will need another person and an area that has some space. Place an object on the floor and call it the sun. If you have three people, you can make one person the sun. Have someone be the earth and walk in a circle around the sun. Now you be the moon and walk in a circle around the earth while the earth circles the sun. The same side of the moon always faces the earth, so while you are the moon, be certain that you are always facing the person that is the earth. Do this several times until you can do it easily.

Was that hard to do? That's what the planets are doing every day, all day long!

Every time a planet goes all the way around the sun and comes back to the same spot, it has completed one revolution around the sun. We would say it has done one orbit, or has revolved one time. When the earth completes one revolution around the sun, we say that a year has passed.

Did you know that as the planets orbit around the sun, they are also spinning around at the same time? This is called rotating. They are rotating as they orbit the sun. Now try this little adventure: just as before, put something or someone in the middle of the floor to be the sun. Now you pretend that you are a planet. Before you begin to walk around the sun, start turning around and around in place. Just twirl around in place, and then begin to walk around the sun while you twirl around at the same time.

I bet you got a little dizzy. Even so, that is what's happening with all the planets every moment of every day.

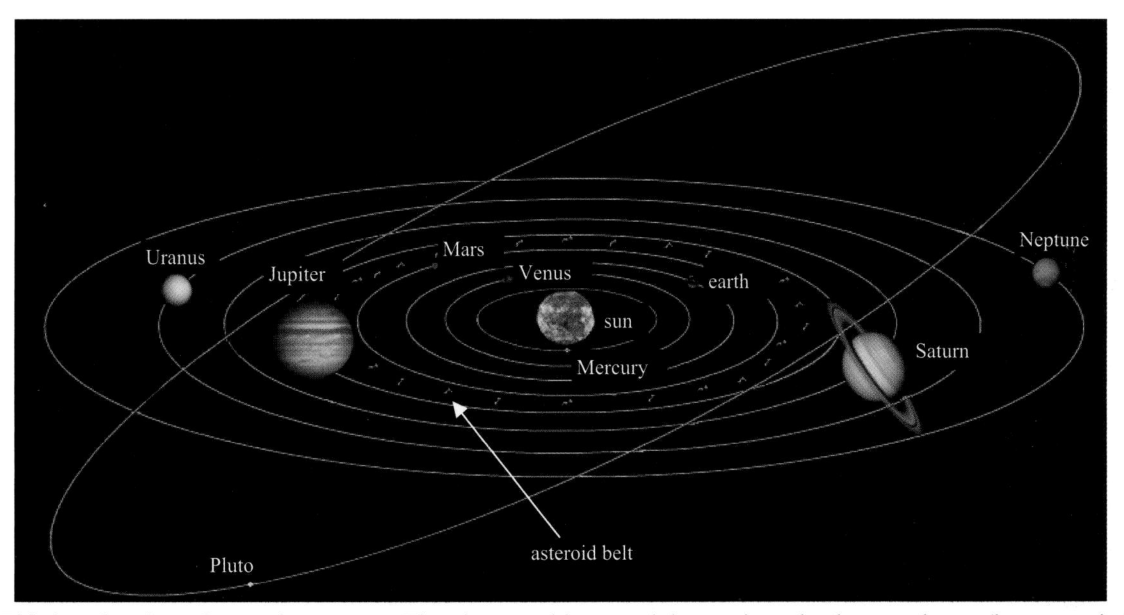

This is a drawing of our solar system. The planets orbit around the sun in paths that are almost (but not quite) circles. While they orbit around the sun, they also rotate. A planet's rotation causes night to turn into day.

Did you notice that while you were twirling around, sometimes you faced the sun and sometimes you faced away from the sun? Well, it's the same with the planets! One part is facing the sun, and then it turns and is facing away from the sun. When we face the sun, it's very bright and light. We call this day. When we are facing away from the sun, there isn't any light shining on us. We call it night. One side of the planet is always facing the sun, so it is always day somewhere on the earth. Every planet rotates, so every planet has a day and a night. Even the moon has a day and a night.

This spinning around, or rotating, is happening right now! We are rotating while we revolve around the sun, and God made it so that we don't even get dizzy! Try not to confuse these two terms: revolve and rotate. Rotating is the spinning that makes it day and night. You can think of it this way: "It rotates between day and night." Revolving is the orbiting around the sun that takes a year. Remember that people say someone's life revolves around things. Well, it's easy to remember that someone's life would revolve around that thing for a whole year.

The earth *rotates*, which turns night into day.

Can you explain in your own words the difference between rotating and revolving? Explain it to someone else today so that you will always remember it.

Solar Flares and Sun Spots

The sun is very active. The fire on the sun is jumping and hopping and rolling about, something like a campfire does. It is a huge ball of very active fire that is spinning around and around. In the large picture of the sun on the right, you can see flares of fire darting out from the sun. That is called a solar flare. It is a giant tower of fire that is many times larger than any planet in our solar system. Look at the drawing of the earth that has been placed in the picture. See how small the earth is compared to the solar flare? That gives you some idea of how big it is! Solar flares burst out millions of miles from the sun. The solar flares throw so much energy and electricity toward the earth that people in the far north and the far south can see colorful electrical lights up in the sky at night caused by these flares. These lights are called **auroras** (uh roar' uhs). Sometimes a solar flare can make a burst of energy come through a city's power lines so forcefully that the entire city can lose its power and electricity. Solar flares can also cause problems in telephones and radios.

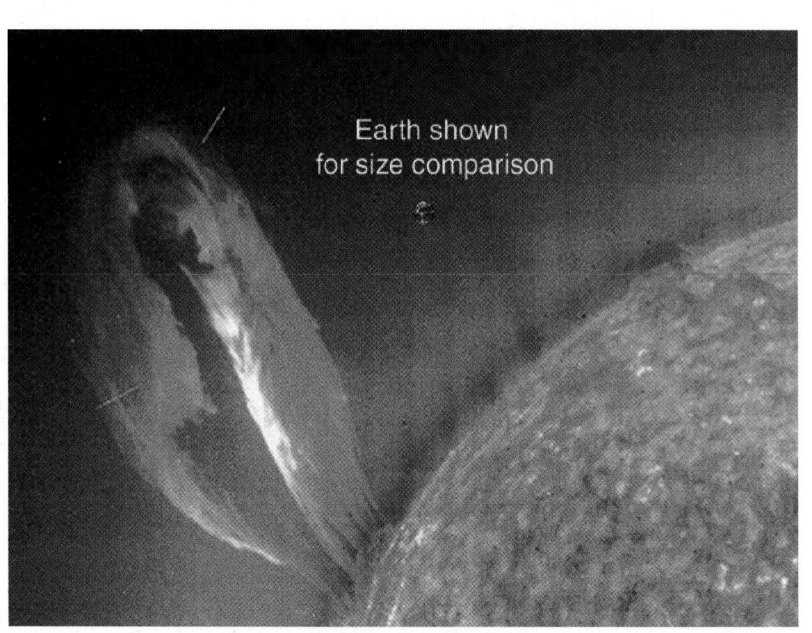

This is a photograph of a solar flare coming off the sun. The earth has been drawn in the picture to give you an idea of how big the flare is.

Look at this picture of the sun. It was taken with special tools used to study the sun. Do you notice anything interesting about this picture? There are little dark spots on the sun. They are called **sunspots**. They are spots that are cooler than the rest of the sun. They are about 4,500 degrees. That's still very hot, but much cooler than the rest of the outer part of the sun. Do you see the dark spot pointed out in the photograph? That spot is bigger than the entire earth! Some are so big that they are ten times bigger than the earth.

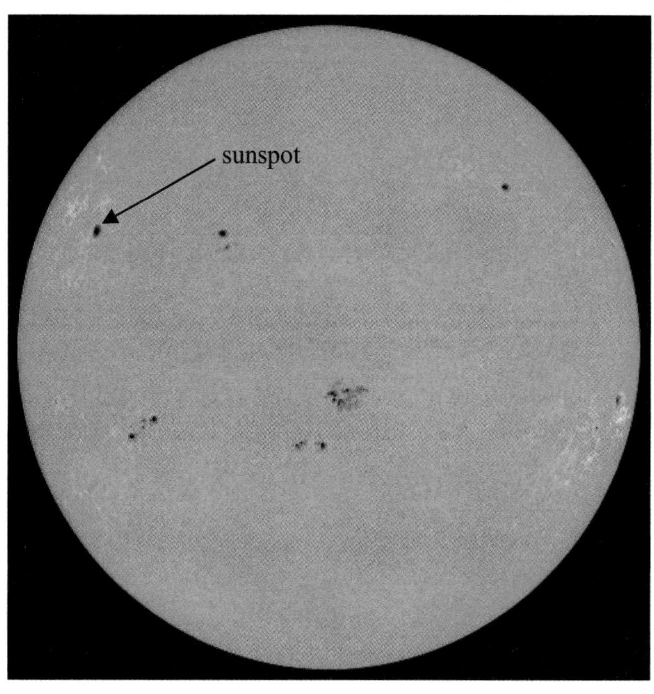

This is a picture of the sun. The dark spots are sunspots.

Many scientists believe that sunspots affect the weather here on earth. You see, when the number of sunspots on the sun is large, the sun is actually *hotter* than when the number of sunspots is small. This may seem strange to you, since the sunspots themselves are cooler than the rest of the sun. However, scientists think that sunspots result in more activity in the sun, so the sun actually gets hotter when there are more sunspots. Without sunspots the earth would be cooler than it is today. This happened a few hundred years ago (1645-1710). During those years, scientists could not see any sunspots on the sun, and the earth was many degrees cooler than normal. In the same way, if there were too many sunspots, the earth would get really hot, and rain would not fall as often. This would cause severe **droughts** (drouts) on the earth. Droughts are long seasons with no rain. When it doesn't rain, the plants die. Without the plants, many animals die as well. People also depend on rain to live. The food we eat cannot grow without a lot of rain. A drought can be a very dangerous thing for life on earth. God has designed our sun with sunspots to help the earth stay at the right temperature. Everything in the universe is God's design. He has taken special care with every feature of the universe to protect us on earth.

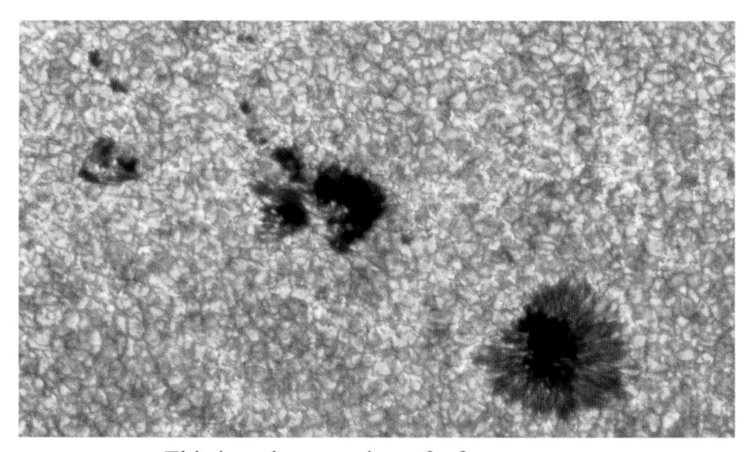

This is a close-up view of a few sunspots.

Mid-lesson Assignments

Because there are so many interesting things to learn about the sun, we are going to break up this lesson with two assignments and an activity.

First Assignment:

Make an illustration for the sun chapter of your notebook.

Second Assignment:

Write or dictate a speech about why people should not look at the sun. You may want to begin with some interesting facts about the sun that you have learned so far. Be sure to explain what happens to your retina when you do it. You can also make a poster with illustrations. Practice your speech several times. Try to memorize it so you will not sound like you are reading when you give your speech to others. It often helps to write one important key word from each sentence on an index card. Then you just glance at the key word and remember what your next sentence is. This will help you to sound like you are *talking* to your audience and not *reading* to your audience. Once you have practiced the speech and feel as though you know it by heart, give your speech to your family, your school, or your homeschool group. If there are several ages of children giving speeches in one family, allow the youngest children to go first.

Mid-lesson Activity

Use a Magnifying Glass to Focus Heat

You will need:

- A magnifying glass
- A refrigerated chocolate bar (Any candy bar that has a lot of chocolate in it will do. Make sure it has been in the refrigerator for a while so that it is cold.)
- A nice sunny day

Instructions

1. Break the chocolate bar into at least three pieces.
2. Take those pieces and your magnifying glass outside.
3. Set the pieces of chocolate bar on the ground, and hold the magnifying glass in your hand.
4. Point the magnifying glass down at the ground away from the pieces of chocolate. Make sure that light from the sun is hitting the magnifying glass.
5. Move the magnifying glass up and down, and play with how you are holding it, until you see a circle of light on the ground. This circle of light is made because light rays from the sun are traveling through the lens of the magnifying glass and are concentrated into a circle.
6. Move the magnifying glass up and down. You should see the size of the spot of light change. It will get bigger or smaller, depending on how close the magnifying glass is to the ground. Get used to how the size of the spot varies as you move the magnifying glass up and down.
7. Lay your pieces of chocolate out on the ground so that they are a few inches apart from one another.
8. Use the magnifying glass to make a spot of light hit one of the pieces of chocolate. Try to make the spot of light as small as possible. Watch what happens for the next few minutes.
9. Do the same thing to the next piece of chocolate, but this time, make the spot of light a little bigger. Once again, watch what happens for a while.
10. Try this on the last piece of chocolate as well, but make the spot even bigger this time. Once again, watch what happens for a while.

What did you see in the experiment? You should have noticed that the chocolate didn't melt much when it just sat out in the sun. However, as soon as you used the magnifying glass, the chocolate should have melted quickly. Why? The magnifying glass concentrated the energy of the sun's light into a small spot. That heated up the chocolate, which made it melt. You should also have noticed that the smaller the spot, the faster the chocolate melted. That's because the smaller the spot, the more concentrated the energy. That resulted in even more heat, which melted the chocolate more quickly.

Creation Confirmation

Did you know that God designed the sun to get its power by little explosions that happen over and over again deep inside the sun? Something called **thermonuclear** (thur' moh new' klee ur) **fusion** is making all those little explosions. That's a big word, but now you know it. So next time someone asks you how the sun gets its power, tell them, "thermonuclear fusion!" What is even more exciting is that thermonuclear fusion makes the sun brighter and brighter each year. Can you believe the sun is actually getting hotter and hotter as it gets brighter and brighter? It is!

Thermonuclear fusion tells us that there could not have been life on earth billions, or even millions, of years ago. You see, since the sun is getting brighter and brighter each year, if we were to go back in time, we would see the sun getting dimmer and dimmer each year. In fact, if we were to go back billions of years, the sun would have been so dim, or faint, that it could not have provided enough warmth for life on earth.

If temperatures on earth were much cooler than they are now, there would be terrible consequences. Oceans would freeze, and it would be winter all the time. Even with the sun's current hot temperature, Antarctica is still very, very cold. Scientists have discovered that the sun would have been many times cooler, actually more than 30% cooler, if it were here billions of years ago. No life could have survived on the earth if the sun were that cool, because the earth would have been a frozen chunk of ice water, with frozen land scattered about. Not a single thing could survive such temperatures. This gives us good evidence that life on earth is young, certainly not millions or billions of years old, as some might want you to believe.

The Color of God's Love

What color is the sun? When we look through special equipment or when the sun is setting in the sky, we see it as orange. However, the color of the sun is actually all the colors of the rainbow! So the sun is red, orange, yellow, green, blue, indigo, and violet. All those colors make every single color in the whole world, so we have a very colorful sun. All the color you see in the world comes from the sun. Without the sun, we would have no color at all!

The sun shines out white light, but that white light is actually all the colors put together, except for black. We call it white light, but it's really very colorful light, since it makes all the colors in the rainbow.

Light always travels in a straight line. It does not bend or go around corners. You know this because when you put your hand in front of your face to shield your eyes from a bright light, like the sun, it casts a shadow across your face. The light does not bend and go around your hand into your face. The reason it is shady under a tree is because light travels in a straight line that does not bend around tree branches.

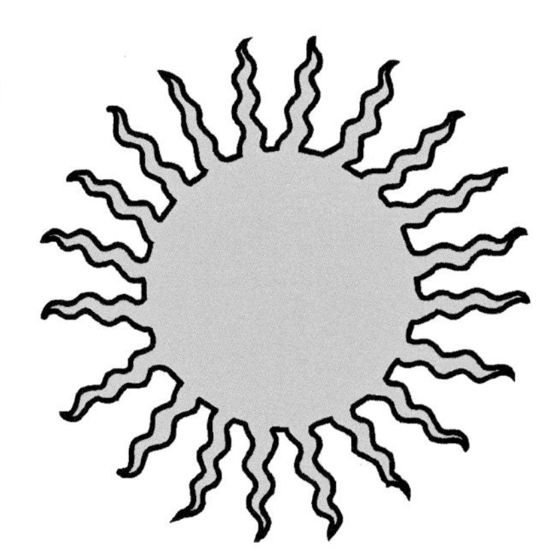

How do we get color? Let me see if I can explain this so that you will understand. Pay close attention. Light is a form of energy. Light energy travels out from the sun in a straight line. Although that line is going on a straight path, the line is curvy, or wavy. In other words, light energy is wavy. Each color is a different size wave. Blue waves are short, as in the picture below. Yellow waves are long, as in the picture below that. I know it's strange to think of color as different kinds of waves, but that is what it really is.

BLUE LIGHT HAS SHORT WAVES

YELLOW LIGHT HAS LONG WAVES

How does it all work? Well, when light hits an object, many of the waves of light absorb into it. To "absorb" means to "take in" or "suck up," the way a towel absorbs water. Almost everything absorbs at least some light waves. Some waves absorb but some don't. Instead, they bounce off. The waves that bounce off the object bounce up into your eye. Your eye sees the bounced light waves. So, when you see a yellow object, you are really seeing the long yellow light waves that bounced off the object and hopped into your eye.

Your eyes and brain are what make it appear yellow. God made your eyes and brain in such a marvelous way. Your eye tells your brain what kind of light it is seeing, and your brain makes the color appear as it does. Now, if there was no eye there to see the object, yellow light would still bounce off the object, but it wouldn't go into anyone's eye to see it as yellow. It would just bounce up until it hit an object that would absorb it. So, here's a question: Is the yellow object still the color yellow if there is no eye there to see it? Hmmm…that is an interesting question. You decide.

Now why does the sun look orange if its light is really all colors? Well, the earth is covered with a layer of mist and gases called our **atmosphere** (at' muh sfear). When the sun's light waves travel through the atmosphere, the short-waved light bounces off the stuff in the atmosphere. When we look up at the sky, those bounced light waves hit our eyes, so our atmosphere appears the color of the bounced light waves. Remember that blue light has short waves. That is why the sky looks blue. All the blue light gets bounced off the stuff in the atmosphere, so the atmosphere looks blue.

If we look at the sun, we see light that has traveled straight from the sun to our eyes. In other words, we see the light that does not get bounced around. If the sun's rays were not bouncing off stuff in our atmosphere, the sun would look white, because all the colors combined make white light. Sometimes the sun does look white when it is straight overhead. That's because the light is not passing through as much atmosphere when it is straight overhead, so not much light is bounced around. As a result, we see all colors of light coming from the sun, which makes the sun look white.

When you see the sun at sunset, the light waves travel through much more atmosphere, and most of the short-waved blue and green light has bounced off by the time it reaches your eye. This makes the sun looks a very deep orange, because those are the colors that do not get bounced around in the atmosphere. We call this a beautiful sunset.

White light coming out of the sun is all the colors together except for black. What, then, is black? Well, when all the colors of light get absorbed into the same object, it makes the object appear black to the human eye. None of the colors bounced off, and that is why it is black. So a black object does not have any waves bouncing into your eye. It absorbs all the energy that the sun's light is pouring down on it. Nothing is bouncing up to make it a particular color. That is why black is not really a color. Color is what you see when something bounces into your eye.

Now, if every single color from the rainbow bounced off the object and into your eye, what color would the object be? Well, think about it. All the colors from the rainbow make white light. So…if the whole rainbow bounced into your eye, you would see the object as white!

White objects do not absorb much light energy, because all colors of light bounce off them. Black objects absorb a great deal of light energy, because they absorb all colors of light. That is fine inside when the light is coming from light bulbs. But when the light is coming from the sun outside, you will notice that black objects are very, very hot. They have a lot of energy sucked into them. White objects are not very hot. They reflect, or bounce off, all the sun's light. So, when you are barefoot in a parking lot during the middle of the summer, and you are waiting for your mom to open the car door, stand on the white dividing lines instead of the black pavement. They are much cooler.

Do you understand why we see color? This is something that is easy to forget. Explaining what you have just learned to someone will help you remember it. Do that now!

God's Light Shines Brighter

The sun is splendid, indeed. But we know God is even more splendid. If it is dangerous to look at the sun because it's so bright, think how much more difficult it would be to look at God in all His glory.

Did you know that when Moses asked to see God, God said, "You cannot see My face, for no man can see Me and live." (Exodus 33:20) So God hid Moses behind some rocks and passed by Moses, who just got a glance of the backside of God's glory passing by. Just from glancing at the tail end of God's glory, Moses' face glowed for a long time. He had to wear a veil over his face so that the Israelites would not be afraid of him. When we read in the Bible about people seeing angels or Jesus when he was transfigured, they are described as being very bright, like a light. God's holy angels are called angels of light.

This is an illustration of the angel Gabriel announcing to Zacharias that his wife is pregnant with John the Baptist. (Luke 1:11-20)

The sun gives us a small clue as to the splendor of God. Isn't it astonishing that our God is so much brighter and more glorious than even the very bright and amazing sun? It's hard for our minds to imagine that, but if we ask God, He will give us a better understanding of all this.

The Bible tells us we will not always need the sun as we do now. When the Lord returns and transforms the world back to its original perfection, God will be the light for us in the day and the night. His light will be all the light we need. Isaiah 60:19 says, "No longer will you have the sun for light by day, Nor for brightness will the moon give you light; But you will have the LORD for an everlasting light, And your God for your glory."

So, even though the sun is so important to life here today, it is only temporary. God is eternal, which means He is forever. We can be certain in our hearts that everything is going to work out according to His wonderful plan! The day will come when God sets up His kingdom on a perfect earth with flawless beauty and splendor in the heavens. We wait eagerly for that day!

Solar Eclipse

What would you think if you went outside one day, in the middle of the day, and there was no sun? You looked around for it, but it was completely gone. It was the middle of the day, but there was darkness all around. Well, something like this happens every once in a while. The sun seems to disappear in the middle of the day for a few minutes, and it seems like nighttime. Then the sun reappears. What makes the sun seem to disappear? It's the moon! The sun gets hidden behind the moon, and its light does not shine down upon us for a moment. It's called a **solar eclipse** (ee klips'). Every year at least two solar eclipses occur. Sometimes there are up to five in one year! You do not see this many, because they are not all visible from where you live. Only certain parts of the world can see an eclipse when it happens. Most eclipses last for only a few minutes, but some have lasted for up to seven minutes. When that happens, animals get confused and sometimes prepare to go to sleep, thinking it's nighttime.

This is a picture of a solar eclipse. The dark circle in the middle is the moon, and you can see that it is blocking out most of the light that is coming from the sun.

A solar eclipse happens when the moon gets right in the path between the earth and the sun. Since the sun's light does not bend around the moon, it casts the moon's shadow across the earth. The moon is teeny tiny compared to the sun, but since the moon is so much closer, it appears bigger and blocks out most of the light from the sun for a few minutes.

Try this to see how it works: Look at the light in the room you are in. Now, close one eye and hold your thumb up in front of your face. When you do this, your thumb blocks out the light. Your thumb is much smaller than the light bulb, but you could block out the light because the light was far away and your thumb was close. Do you see how the moon could make the sun seem to disappear?

When there is a **total eclipse**, the sun is completely hidden behind the moon. During an **annular** (ah' yuh ler) **eclipse** the sun is directly behind the moon, but a ring of sunlight can be seen around the blackened moon. This happens because the moon is sometimes farther away from the earth during an eclipse, and it cannot completely hide the sun.

With one eye closed, hold your thumb so it blocks the light in your room again. Now slowly move your thumb farther away from you. Can you see a little bit of the light bulb behind your thumb? That is like an annular eclipse.

Sometimes during an eclipse, the moon is not directly between the sun and the earth. It is only partially between the two. These are called **partial eclipses**. "Partial" means, "part." During a partial eclipse, you can see a part of the moon pass in front of the sun. It is still an amazing site to see. Full and partial eclipses are astounding, but you must only look at them through special eclipse-viewing glasses or a special eclipse-viewing box. It is important that you understand that looking directly at the sun, even during an eclipse, could cause permanent eye damage and even blindness. We will make a special eclipse box at the end of this lesson so that you can view the sun and a solar eclipse safely. You can save the box for when there is an eclipse in your area.

The moon is not a perfectly round ball. It has many gigantic holes, called craters that give it an uneven surface. You can see these holes very clearly through a telescope. During an eclipse, little points of light reflect off these holes. We call these bright points of light **Bailey's Beads**. They look like little bright balls, or beads.

As long as you use the right equipment, watching a solar eclipse is a lot of fun. Of course, solar eclipses don't happen very often, so you need to make sure you know when the next one is. That way, you won't miss it. Go to the course website that I told you about in the introduction to the course. That website will have links to other websites that tell you when the next solar eclipses are supposed to happen. Make sure to mark them on your calendar so that you won't forget to watch!

What Do You Remember?

Do you remember how many earths would fit inside the sun? How many miles away is the sun? What is the solar system? Explain what sunspots are. Do sunspots help us at all? Does the sun have a satellite? Can you explain the difference between revolving and rotating? How does the sun tell us that there were not living things on the earth billions of years ago? Do you remember why you see color? Which color has short waves? Can you explain what a solar eclipse is? If you had trouble answering these questions, it might be a good idea to read this lesson again. Explaining these things to someone will help you better understand and remember what you have learned.

Assignment

Older Students: Look through old magazines and cut out pictures of the sun. Make a collage with all these pictures in your notebook. On a separate page of your notebook, write down important facts that you wish to remember about the sun.

Younger Students: Have your child look through old magazines to cut out pictures of the sun. Help him to make a collage of these images in his notebook. Have him dictate to you what he remembers about the sun, and write that in his notebook.

Activity

Make a Solar Eclipse

You will need:

- A flashlight
- A globe (or round ball)
- A Styrofoam® ball (or any small ball) on a string

1. Place your globe on a table or floor.
2. Set your flashlight upon a stack of books until it is shining directly on the center countries of your globe.
3. Turn off the lights.
4. Lower the Styrofoam® ball down between the flashlight and the globe, but make sure it is closer to the globe than it is to the flashlight. Move it until it is projecting a dark shadow on the globe, as shown in the picture. Notice the dark, round circle that is projected onto the globe. That is called the "umbra" (uhm' bruh) of the eclipse. The people within that dark circle experience a total solar eclipse. The lighter shadow that surrounds the dark circle is called the "penumbra" (puh nuhm' bruh), and the people in that section experience a partial solar eclipse. Notice that the umbra and penumbra are not on most of the globe. The parts of the globe that do not have an umbra or penumbra do not experience any kind of eclipse. This is why a solar eclipse is not visible in all parts of the world. Though we have several eclipses a year, they are visible in only a few countries.

Project
Pinhole Viewing Box

You are going to make a pinhole viewing box that will help you to safely study the sun. You will not look at the sun directly through the pinhole. Instead, you will project an image of the sun onto a sheet of paper at the back of your box. The light from the sun will come through your pinhole and will shine an image of the sun onto the paper.

You can use this to view the sun and, if they are big enough this year, you will be able to see its sunspots! Be sure to save it for when there is a solar eclipse! **Do not ever look through the pinhole directly at the sun**. The pinhole will shine the shape of the sun onto the back of your box, and that's what you need to look at.

Instructions

You will need:

- A parent
- A box
- Scissors
- White paper
- A pin or needle
- Tape
- Aluminum foil

1. Find a box. The length of the box is important. The longer the box, the bigger, but fuzzier, the pinhole image. A shorter box gives you a smaller, clearer image. If you can't find a long box, you can tape together two or more boxes to make a longer one.
2. Use the scissors to cut a viewing hole in the side of the box. You can cut the whole side off of the box, or just a hole big enough so that you can see the back end of the box.
3. Use the scissors to cut a hole in the center of one end of the box. This will be the pinhole side.
4. Tape a piece of foil over the hole you just cut into the pinhole side.
5. Use the pin to poke a small hole in the foil. Make sure that the pinhole is over the hole that you cut in step 3, so light that passes through the pinhole will reach the other end of the box. You will not be looking through this pinhole; it is for the sun to shine through onto the other side of the box. A small hole will give you a sharp, but dim image. A larger hole will give you a brighter, but fuzzy image.

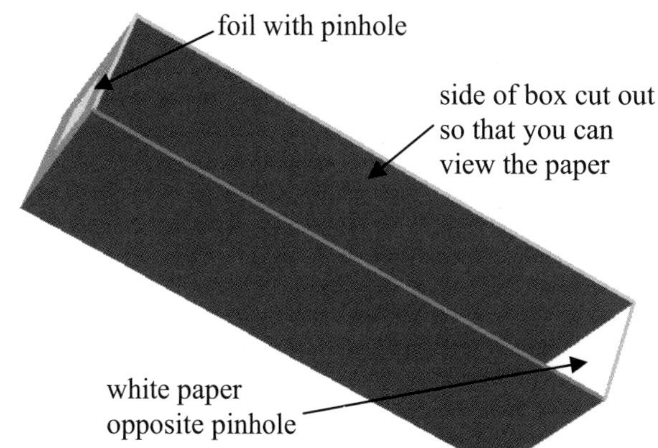

6. Put a piece of white paper inside the back end of the box, opposite the pinhole.

7. Point the end of the box with the pinhole at the sun so that you see a round image on the paper at the other end. If you are having trouble pointing the box at the sun, look at the shadow of the box on the ground. Move the box so that the shadow looks like the end of the box. In other words, make sure that the sides of the box are not casting a shadow. The round spot of light you see on the paper is a pinhole image of the sun.
8. As you look at the image of the sun, see if you find dark spots within the image. Those are the sunspots! Make an illustration of them in your notebook.
9. If you want a brighter image of the sun, you can remove the foil that you currently have and replace it with new foil. Then, poke a larger hole in the new foil. If you want a sharper (but dimmer) image of the sun, replace the foil with new foil and poke a smaller hole in it.
10. If you want a larger image of the sun, you can make another pinhole viewing box out of a longer box. If you want a sharper (but smaller) image of the sun, you can make another pinhole viewing box out of a shorter box.
11. If you use your viewer during an eclipse, you will be able to safely observe the moon moving over the sun, blocking its light.

Look only at the image on the paper! Do not look through the pinhole at the sun!

sun

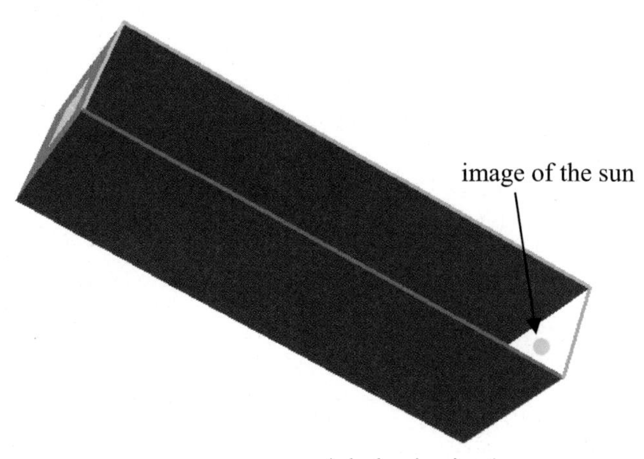

pinhole viewing box

Lesson 3
Mercury

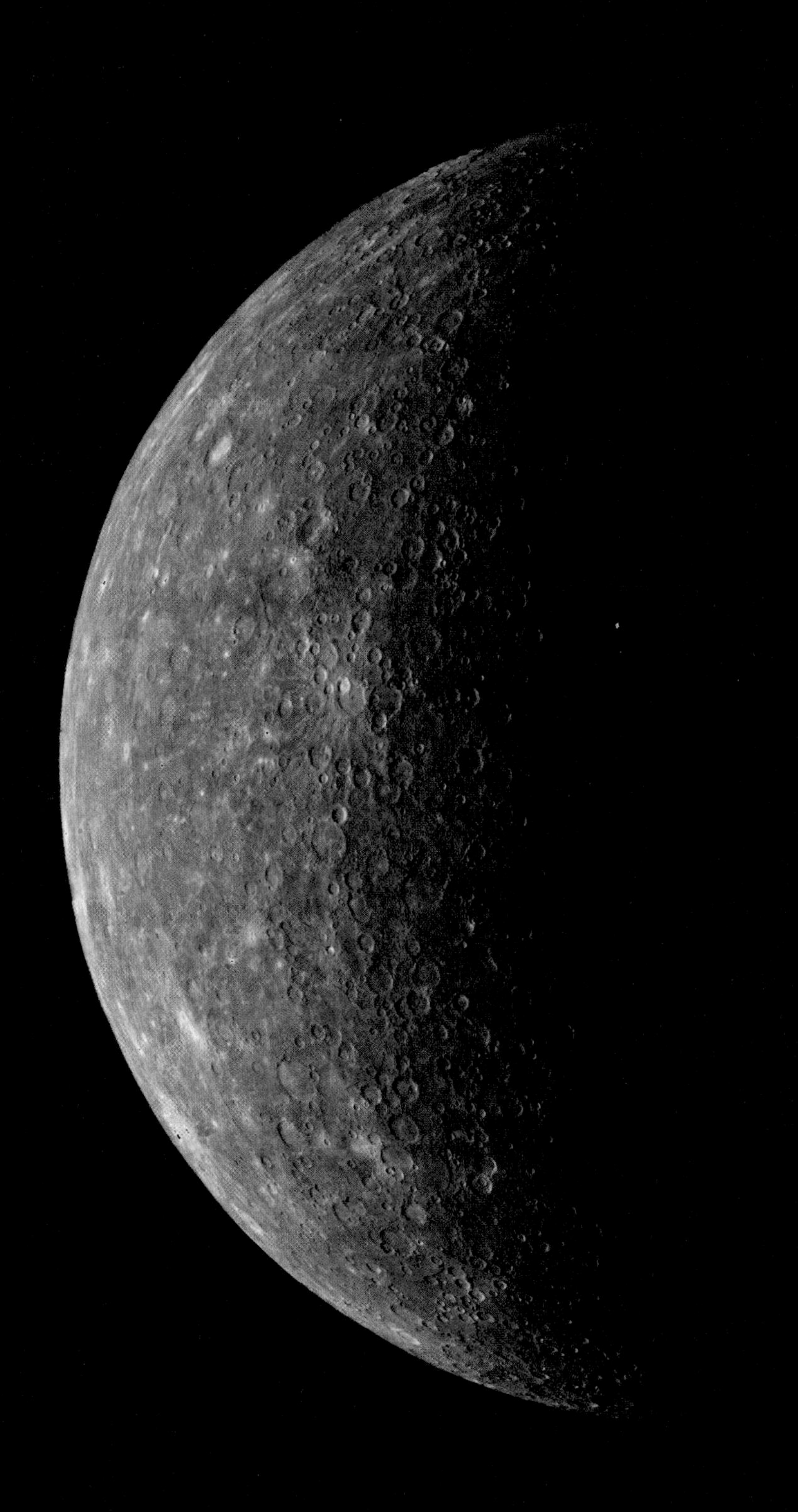

The Planet Closest to the Sun

Because Mercury is the closest planet to the sun, we will study it next. We actually do not know as much about Mercury as we do about many of the other planets in our solar system, because it is so close to the sun that it is very hard to observe through telescopes. The only spacecraft that has gotten close to Mercury is called "Mariner 10." Although Mariner 10 collected a lot of information about Mercury, it is still very little compared to what we know about many of the other planets. Of course, we do know *some* things about Mercury, and you will learn many of those things in this lesson.

This is a picture of Mercury as taken by Mariner 10. It is actually a composite of several individual pictures. The plain brown patches indicate regions that Mariner 10 could not photograph.

If you were standing on Mercury, the sun would seem a lot bigger to you. It would seem gigantic because you would be closer to it. And boy, would it be hot and bright!

During the day, when the sun is shining on Mercury, it gets sizzling: more than 750 degrees. That's hotter than an oven! You might think that since it's so close to the sun, it would never get cold there. You might think that it's always hot, but that is not so. Mercury cools down a lot at night, because the sun does not shine then. So, when it is night, Mercury gets colder than a freezer (-300 degrees). The reason this happens is because Mercury does not have much of an atmosphere. As you learned in Lesson 1, the atmosphere is the layer of mist, clouds, and gases that covers a planet and holds the heat in. The earth's atmosphere (we call it "air") is made up of several gases. You probably know that one of those gases is called "oxygen" and that we breathe it to stay alive. Because Mercury does not have oxygen, we could not breathe on Mercury without a spacesuit.

The side of Mercury facing the sun is bright, sunny, and warm. The side facing away from the sun is dark, cold, and bleak.

Earth's atmosphere holds in the heat from the sun so that the earth stays pretty warm at night. Without an atmosphere to hold the heat in during the night, Mercury gets freezing at night. We would never feel very comfortable on Mercury. We would always complain that we were too hot or too cold.

Because Mercury doesn't have an atmosphere, the sky always looks dark. There are no air particles to scatter the light waves all over the sky. Remember, the reason the sky looks blue on earth is because particles scatter the sun's blue light across the sky. Can you guess what color the sun would be to someone standing on Mercury? The sun would appear white all the time! Can you explain why? Can you explain why Mercury is so cold at night even though it is right next to the sun?

This is an artist's idea of what Mercury might look like. Notice that although the sun is shining, the sky is dark.
Digital artwork by Dr. David Heatley

Rotation and Revolution

Remember how I explained that a planet spins around in place while at the same time moves in a circle around the sun? The spinning in place is called rotating, and moving in a circle around the sun is called revolving. Planets rotate as they revolve. Of course, one part of the planet is always facing the sun, and the other part is always facing away from the sun. The earth takes 24 hours to rotate once. This means that if you are on the side of the earth that is facing the sun, you will be in the same position (facing the sun again) in 24 hours. We call this an "earth day," because it is the length of time that a day lasts on earth. Compared to the earth, Mercury doesn't spin fast at all. It spins very slowly. Mercury takes 59 earth days to rotate one time. A full day on Mercury would be just under 30 earth days of bright, hot sunshine and just under 30 earth days of cold darkness.

I think we would miss the nighttime if we had to live a full day on Mercury, and we would miss the daylight if we had to spend a full night on Mercury. Imagine going to bed at night with that bright, hot sun still shining through your window and waking up with that bright, hot sun still shining on and on for almost 30 earth days. Imagine it the other way, waking up in pitch dark and spending the day in freezing cold darkness, going to bed, only to wake up to freezing cold darkness again for almost 30 earth days! We can see that God did not intend for people to live on Mercury. There is only one planet just perfectly made for humans- earth! Earth gets dark just about the time we feel tired, and about the time the sun is coming up, we are ready to wake up! It's not by accident that the earth is perfectly suited for our sleep cycle. God made the earth's spin just for us.

Even though Mercury is rotating very slowly, it's revolving around the sun very quickly. It's practically racing around the sun. It only takes 88 earth days for Mercury to make one full trip around the sun. That's pretty fast. Mercury is the fastest planet in the solar system. It takes earth a whole year, 365 earth days, to revolve all the way around the sun. Strangely, one day on Mercury is only a little shorter than a year on Mercury. After all, a day on Mercury is 59 earth days, while a year is only 88 earth days. If you lived on Mercury, you would be 33 years old on Mercury when you were only eight years old on earth!

Mercury revolves around the sun in an oval pattern, not in a circle. We call it an **elliptical** (ee lip' tik uhl) orbit. If it traveled in a perfect circle around the sun it would always be the same distance away from the sun. An elliptical orbit makes it sometimes closer to the sun and sometimes farther away. How do you think it affects Mercury to be closer to the sun? How do you think it affects Mercury to be farther away?

Because Mercury's orbit is elliptical, it is sometimes closer to the sun and sometimes farther away from the sun. In this drawing, it is closer to the sun when it is on the right (where it is drawn), and it is farther from the sun when it is on the left.

Actually, not one of the planets orbits the sun in a perfect circle. They all have elliptical orbits. However, most of the other planets (including earth) have orbits that are almost circular. In other words, even though earth is sometimes a little closer to the sun and sometimes a little farther away from the sun, the difference is very, very small. Because of this, we usually call the orbits of most of the other planets circles, even though they are all actually elliptical. Mercury and Pluto have the two most elliptical orbits in our solar system.

Can you explain in your own words all that you have learned about Mercury so far?

Features of the Planet Mercury

Mercury is small. It is much smaller than the earth. If the earth were the size of a baseball, Mercury would be the size of a golf ball. It's about the same size as our moon. In fact, there is only one planet in the solar system that is smaller than Mercury, and that's the planet Pluto. Later on in this course, you will find out that Pluto is so small that there are those who think it should not even be considered a planet. In the minds of those people, Mercury is really the smallest planet in the solar system.

This is a size comparison of the earth and Mercury.

What is Mercury like? It's kind of empty and lonely, like our moon. If you stepped on Mercury you would get dust all over you, because it's very dusty and dry. Mercury has many craters on its surface. A crater is a large dent on the surface of the planet. It is a place where space rocks, called **asteroids** (as' tuh roids), crashed into Mercury, leaving big gigantic dents. If you have ever had a sandbox, it would be like dropping rocks into the sand and then picking them up to see the dents they left. We will make craters in flour at the end of this lesson. Craters are kind of like a scar on the surface of a planet. Do you have any scars? If so, they are not as big as the craters on Mercury, because Mercury's largest crater is as big as the state of Texas! If you stood on the edge and looked down into the crater, it would be a long, long way down to the bottom.

This is a picture of Mercury's surface.

Astronomers believe that asteroids fell out of the sky and crashed into Mercury, leaving all these craters on the surface. The asteroids could have come from a planet that exploded at one point in time. Many believe that there was once a planet between Mars and Jupiter that exploded and sent pieces of it flying into space. The planets and moons that did not have strong atmospheres to protect them would have received a lot scarring if such a thing actually happened. The earth's atmosphere burns asteroids up in the air before they reach our planet. Mercury doesn't have this protection, so if thousands of monstrous-sized rocks flew into Mercury, they would have caused huge pits on its surface.

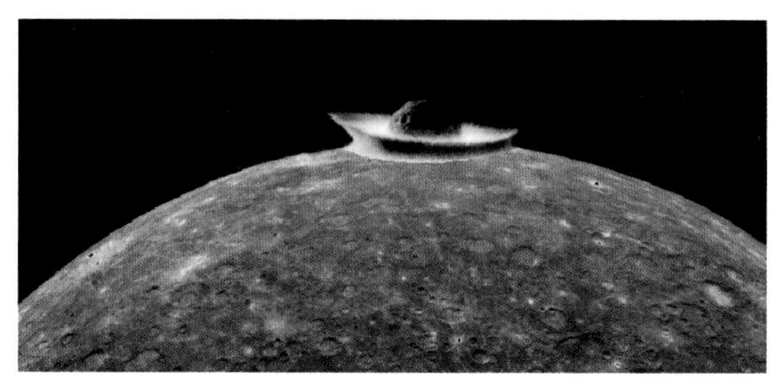

This is an artist's idea of what it might have looked like when a large asteroid crashed into Mercury.

Amazingly, parts of Mercury have no craters. The fact that parts of Mercury are craterless is difficult to understand for those who believe that the solar system is millions or billions of years old. You see, over millions or billions of years, every part of the planet would have gotten hit many, many times by falling asteroids. Scientists know that the chances of some parts of Mercury never getting craters over billions of years is next to impossible. The best explanation for why Mercury has sections with no craters is that the solar system is not millions or billions of years old. However, scientists who want to believe that the solar system is that old have come up with another explanation. They say that the

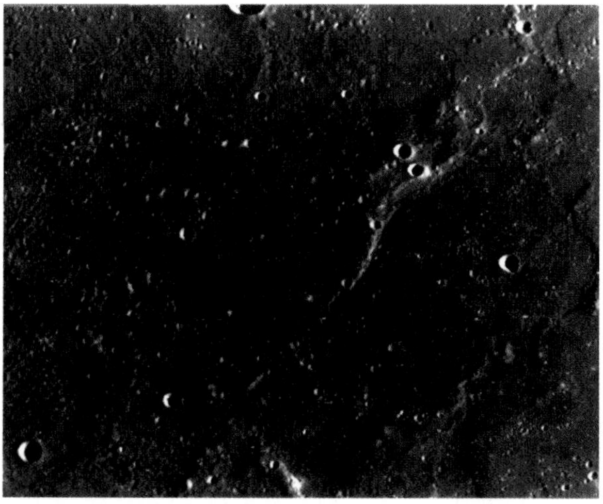

Although most of its surface is pitted with craters, parts of Mercury are almost craterless. Note how few craters there are in this photo.

craterless sections are the "new" parts of Mercury. According to these scientists, the "new" parts of Mercury were formed recently by volcanic eruptions. Since these sections are not very old, they have not had time to be struck by asteroids yet, so they have no craters. Of course, we know that God created the whole planet of Mercury instantly, with only a Word. I also believe that the *whole planet* is not nearly that old, because I think the Bible tells us that God spoke it into existence only a few thousand years ago.

Mercury is a rocky planet. We call it a **terrestrial** (tuh res' tree uhl) planet. Terrestrial means "earth like." Now of course, this doesn't mean that a terrestrial planet is *a lot* like the earth. It just means that the planet is solid. In other words, you can stand on its surface. There are five terrestrial planets in our solar system. The other four planets are (believe it or not) **gaseous** (gas' ee us) planets. This means they are made of gas. In other words, they are not solid; they don't have ground to stand on. You cannot land on a gaseous planet, because you would sink into it.

Scientists believe the inside of Mercury has the same material as the inside of the earth. This would mean that Mercury has an iron and nickel core. The core is the very center of the planet, just like the core of an apple is the center of the apple.

Spacecraft to Mercury

No man has ever gone to Mercury, but we have sent a spacecraft to Mercury to get information for us. Only one spaceship has ever been to Mercury. As I told you at the beginning of the lesson, it was called "Mariner 10." No person was on Mariner 10. We call it an **unmanned spacecraft** because there were no people on it. Even though there were no people on the spacecraft, there were many, many scientific instruments onboard. These instruments allowed the spacecraft to collect all kinds of interesting information about the planets it visited.

This is an artist's idea of what Mariner 10 looked like as it traveled through space.

Mariner 10 was launched in November of 1973, and its mission was to take pictures of and collect information about Venus and Mercury as it flew past them. Mariner 10 reached Venus in February of 1974. After scientists used Mariner 10's instruments to study Venus, they sent the spacecraft signals that caused it to head to Mercury. It reached Mercury in March of 1974, and it was able to fly past Mercury again in September of 1974 and again in March of 1975. Most of the information we have about Mercury came from Mariner 10.

How old was your mother in 1974? Ask your mom what year she was born, then do a subtraction problem. Take the year your mother was born and subtract it from 1974. That will tell you how old she was when Mariner 10 first visited Mercury. For example, if your mom was born in 1968, you would do the problem like this:

$$\begin{array}{r} 1974 \\ -\underline{1968} \end{array}$$

6 years old!

Although no spacecraft has visited Mercury since 1975, NASA plans for another spacecraft (called MESSENGER) to visit Mercury in 2009. You can learn more about this new NASA mission to Mercury by going to the course website I told you about in the introduction.

A Trip Across the Sun

Because Mercury is closer to the sun than we are, it can sometimes come between the earth and the sun. As you learned in Lesson 2, when the moon comes between the earth and the sun, we call it a solar eclipse, and the moon can actually block out the sun. Well, since Mercury is much farther away from us than the moon, it cannot block out the sun. However, it can block out a tiny portion of the sun, forming a small, black dot. That's what the picture on the right shows. The tiny black dot pointed out by the black arrow in the picture is there because Mercury is between the earth and the sun, blocking out that portion of the sun's light. When astronomers see this, they say that Mercury is "transiting the sun," because "transit" means "to pass over."

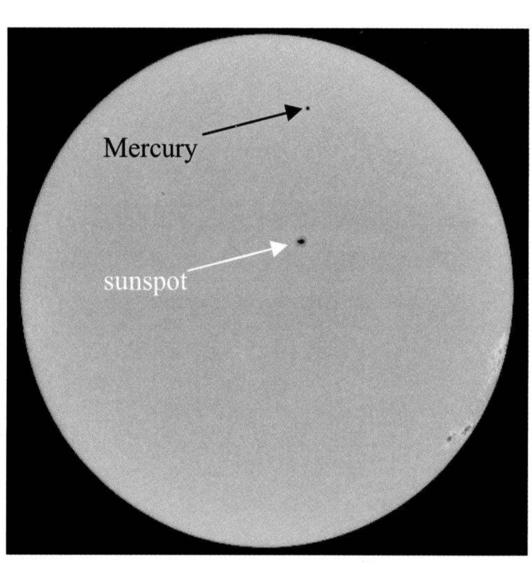

This is a photograph of the sun during the Mercury transit of May 7, 2003.

Because Mercury is traveling around the sun, if you were to watch a transit of Mercury, you would see the black dot move across the sun. Scientists took several photographs of the sun during a recent Mercury transit, and then they compiled them into a composite (kahm pahz' it) picture that shows the position of Mercury at different times during the transit. The word "composite" means "made up of many parts." A composite picture, then, is made up of many pictures. Look at the "trail" of black dots in the picture. It shows you the path Mercury took as it traveled across the sun.

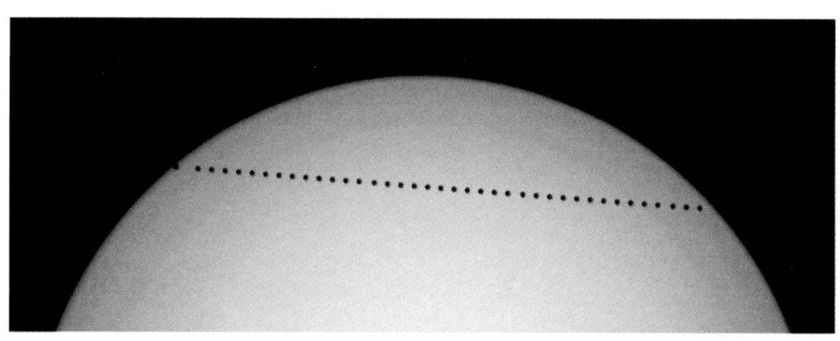
This is a composite photograph of the upper portion of the sun during the May 7, 2003 Mercury transit. It shows you how Mercury traveled in front of the sun during the transit.

Who Named Mercury?

Sadly, the Romans who first named Mercury did not know God. Jesus had not come to save the people of the world from their sins yet. The Romans worshipped idols and named the stars in the sky after these idols. The Romans noticed that Mercury moved in the sky very quickly as compared to the other planets. After all, Mercury's orbit is so fast that it travels around the sun four times in just one earth year! Because of this, the Romans named Mercury after the god of travel. You will be happy to know that when Jesus came, many Romans gave up their false gods and became Christians. Eventually many of the people in the city of Rome became followers of Christ, and Christianity became the major religion there. It's too late to rename the planet something else, but we can be happy that Christianity is still the major religion in that part of the world today.

How to Find Mercury in the Sky

Did you know you can see Mercury without a telescope? You can see it either just before the sun comes up in the morning or just after the sun goes down in the evening, but not in the middle of the night. It's so close to the sun that we can't see it at night, when we are facing away from the sun. If you look carefully in the sky, toward the rising or setting sun, you may just spot this special planet very close to the horizon. It looks like a star because the sun's light shines upon it and makes it light up. Even though it shines like a star, there is really no light coming from Mercury. The light that we see is just the sunlight reflected off Mercury's surface. Mercury is hard to find because it is often hidden from our view by the glare (bright rays of light) from the sun.

Even though Mercury is very hard to find the sky, you might want to try. You can get some help finding Mercury by going to the course website that I told you about in the introduction. The website will have links to places that will show you where Mercury can be found if it is visible when you are looking for it.

What Do You Remember?

Let's see what you remember about Mercury. How long is a day on Mercury? How long is a year on Mercury? Which is longer, a day or a year? Does Mercury orbit in a circle or in an oval around the sun? What is the shape of Mercury's orbit called? Is it hot or cold on Mercury? Why is it so cold at night? What kind of planet is Mercury: terrestrial or gaseous? What does the surface of Mercury look like? What are some reasons it might look like this? What would the sky look like if you were on Mercury? Why? When is the best time to see Mercury? Why?

Assignment

Now it's time to record what you have learned about Mercury. Make an illustration of Mercury for your notebook. You might want to try to illustrate the oval shaped orbit of Mercury traveling around the sun. Write down the interesting information you learned about Mercury in your notebook.

Activity

To help you understand more about craters, we are going to make craters in a bowl of flour.

You will need:

- ♦ A parent
- ♦ A large bowl
- ♦ Flour
- ♦ Several pebbles of different size

1. The pebbles you collected will represent asteroids in this activity. You can use marbles if you like, but use some pebbles, because most asteroids are not perfectly round like marbles.
2. Fill a bowl halfway with flour.
3. Place the bowl on the floor and drop the smallest pebble into the bowl.
4. Very carefully, trying not to disturb the shape of the crater, pick the rock pebble. Take note of the size of the crater it made.
5. Continue to drop pebbles of different sizes into the bowl. Each time, remove the pebble carefully and note the kind of shape that was made in the flour.

Do you see anything special about the flour shapes that you made? What kinds of shapes were made when you dropped pebbles that were not perfectly round into the bowl? What did the crater look like? You should see that even pebbles that are not perfectly round make craters that are almost perfectly round. Isn't that interesting?

Project
Make a Model of Mercury

For this project, you are going to make a model of Mercury with flour and salt. It is called a salt dough model and it's very easy.

You will need:

- ♦ A parent
- ♦ A large bowl
- ♦ A marble or pebble
- ♦ A pencil
- ♦ 1 cup of white flour
- ♦ ¼ cup of salt
- ♦ 1 teaspoon cooking oil
- ♦ ¼ cup of water

1. Mix the flour, salt, cooking oil, and water in a large bowl. It will form dough.
2. If the dough doesn't stick together, add a few more drops of water. Do not add too much water, or it will be gooey and sticky and will not dry into a ball.
3. Form the dough into a ball.
4. After you have made the ball of dough, you want to make it look like Mercury. To do this, you will make the craters on the ball. Look at the pictures below to see what your craters should look like:

5. You can use marbles or pebbles and pretend they are asteroids falling onto the planet, making the craters. Since the craters should be different sizes, you should also make some craters with a pencil. Use both sides of the pencil (the tip and the eraser) to make lots of differently-sized craters. Make craters all over your model of Mercury.
6. When you are finished, place your salt dough model of Mercury somewhere to dry.

Lesson 4
Venus

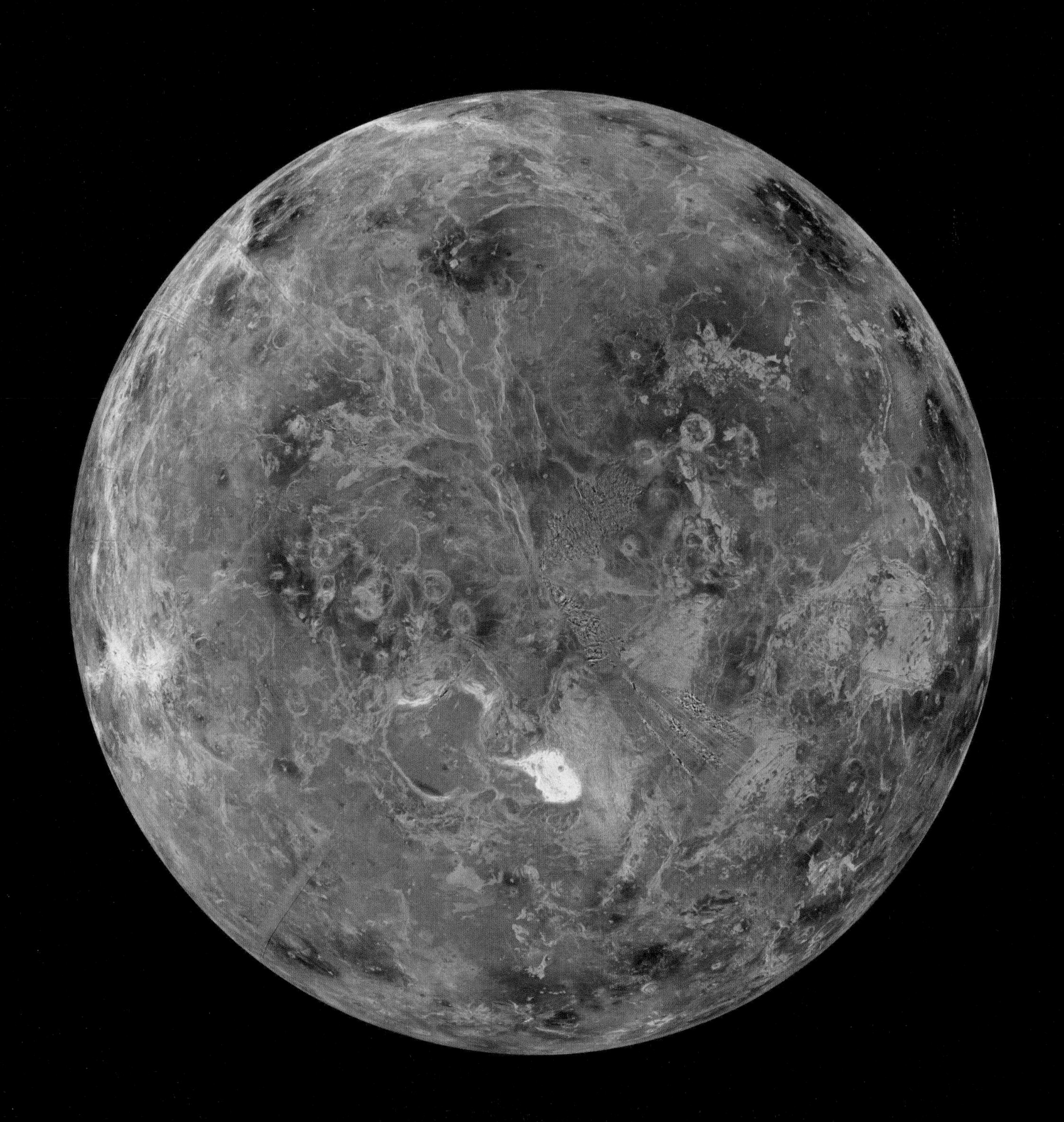

Venus

Hot! Hot! Hot! The hottest planet in the whole solar system is not the planet closest to the sun. It's the one that is second closest. Venus is burning hot and stays hot all day and all night. It is about 870 degrees on Venus. There isn't another planet as hot. It also has lots of volcanoes that spew lava onto its surface. Do you know what lava is? It is rock that is so hot it has melted, like butter melts in a pot on a stove. Lava, however, is much, much hotter than that. It is so hot it would melt the pot holding the butter, turning the pot into liquid! Lava comes out of a volcano when it erupts. When it is inside a volcano, this lava is called "molten rock" (which means "melted rock"). When it comes up to the surface and pours out of the volcano, it is then called lava.

This is an artist's idea of what active volcanoes on Venus might look like. The red liquid in the picture is lava.
Digital artwork by Dr. David Heatley

Venus has thousands of volcanoes all over it. They blow up every now and then, spouting and spraying hot, fiery lava all over the place. Many of the parts of Venus that are not covered with hot lava are covered with places where lava once flowed and then cooled down enough to turn into hard rocks. Even though these hard rocks are no longer lava, they are still very hot, because everything on Venus is hot. Venus's whole surface is affected by these volcanic eruptions. When a volcano erupts here on earth, it wipes out every living thing it touches. Every tree, house, plant or animal that it touches becomes nothing but ash when the lava dries. When a volcano erupts, for years afterward, people are still trying to clean up all the ash that blows around the area.

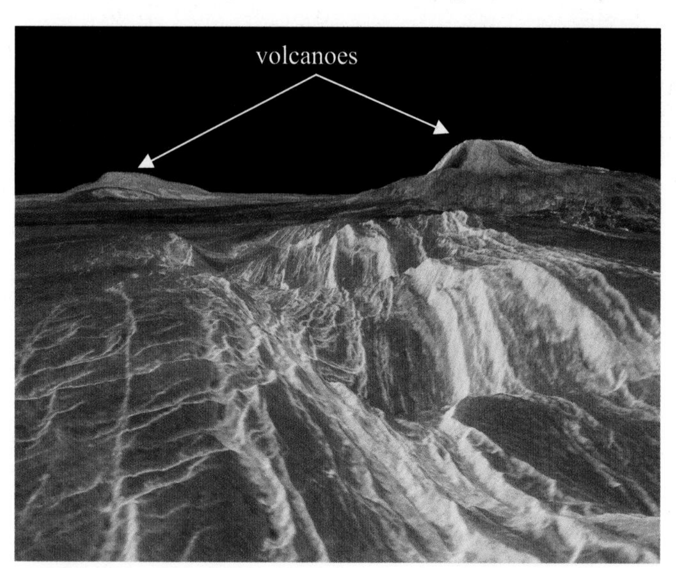

This is an image of part of the surface of Venus. There are two volcanoes in the background, and the lighter-colored rocks are rocks that were once lava but have since cooled to form hard rocks.

Do you have a fireplace? If you have real logs in your fireplace, you have seen the ashes that are left when they have burned up. Venus probably doesn't have ashes all over it, because there is nothing living on Venus, so there is nothing to burn up and turn into ash.

With all these rocks all over Venus, do you think it is a terrestrial planet or a gaseous planet? Remember that "terrestrial" means "earth-like." Yes, Venus is a terrestrial planet. If a planet has a rocky or earthy surface, it is terrestrial, because you can stand on its surface, just like you can stand on earth.

Mid-Lesson Activity

Make Some "Lava"

You will need:

- A parent
- ¼ stick of butter or margarine
- Some flour
- A large plate
- A small bowl (like a custard dish) or a small cup
- A saucepan
- A stove

1. Melt the ¼ stick of butter in the saucepan on a low heat setting. You will probably have to swirl the butter in the saucepan from time to time in order to get it to melt completely.
2. Once the butter is melted, pull the saucepan off the hot burner and turn off the heat.
3. Put the custard dish or small cup upside down in the middle of the large plate.
4. Sprinkle flour around the custard dish or small cup so that you have a nice layer of flour surrounding it. Your plate should look like the picture on the right.

5. Slowly and carefully pour the butter out of the saucepan onto the center of the custard dish or cup. Watch it roll down the sides of the dish or cup and into the flour. As it does so, you should see it form "streams" of melted butter in the flour.
6. Let the plate sit out for a few hours. Do not disturb it.
7. Come back later and look at what has happened.

In this activity, the butter represented rock. It started out solid, but when you heated it, it melted, turning into a liquid. Rocks do the same thing. When heated in the hot insides of a planet, rocks melt. When you poured the liquid butter over the custard dish, it was like lava pouring down the sides of a volcano. It cut little streams out of the flour, going where it wanted to go. After you let the plate sit out for a while, what happened? The butter hardened up again, so that it was no longer a liquid. This is what happens after lava comes out of a volcano and then sits for a while. It hardens into rock.

Too Much Atmosphere

You might be wondering why Venus is hotter than Mercury, which is right next to the sun. There is a very good reason. Do you remember what the word "atmosphere" means? The atmosphere is made up of all the gases that surround a planet. Venus has a very heavy atmosphere, with thick, heavy clouds swirling around it. These clouds and the atmosphere trap the heat that comes from the sun, and the heat cannot escape Venus and go out into space. As a result, the sun's heat comes in, but it doesn't go out very easily. This keeps Venus very hot. The clouds on Venus are made of different stuff from the clouds on earth. They are heat-trapping clouds made of sulfuric (suhl fyur' ik) acid. Sulfuric acid is not good to breathe.

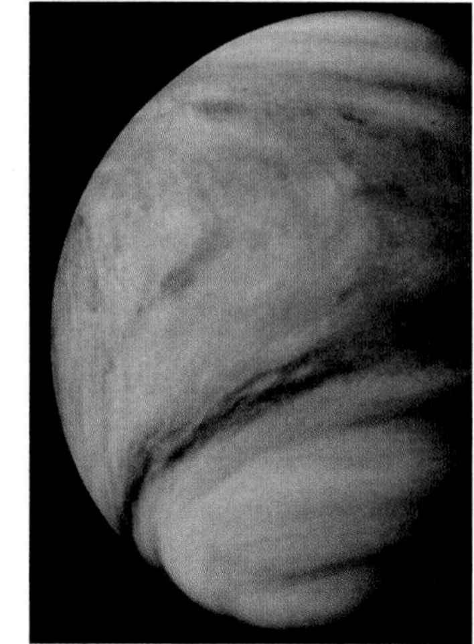

This is a picture of the clouds that constantly surround Venus.

When your mother is trying to make water boil in a pan on the stove, it will boil faster if she puts a lid on the pot. The lid keeps the heat from escaping, trapping it inside the pot. This makes it hotter and hotter inside the pot so that the water will boil more quickly. The clouds on Venus are similar to the lid on the pot; they surround the planet, keeping the heat from escaping into space.

The clouds and atmosphere that trap heat on Venus are always moving across the planet. They move quickly and are always circling the planet, heating up everything they pass over. So, the side of Venus that faces away from the sun isn't cooler than the side that faces the sun. Remember, this is not what it's like on Mercury. On Mercury, the side of the planet that faces away from the sun gets very cold, because there is no atmosphere and no clouds to keep it warm. On Venus, the side that faces away from the sun is kept warm by the atmosphere and clouds, so it never has a chance to cool off. Because of this, it's just as hot at night on Venus as it is during the day.

Rotation and Revolution

Venus rotates slowly. Do you remember what rotate means? It is the circular movement that makes a planet face the sun and then turn away from it, changing day into night. Venus rotates even more slowly than Mercury. It takes 243 earth days for Venus to make a full turn. Because of this, daytime on Venus lasts just over 121 earth days, and nighttime lasts just over 121 earth days.

Another funny thing about Venus is that it rotates in the opposite direction as compared to the earth. On the earth, when we see the sun come up early in the morning, we know we are looking east, because the sun always rises in the east. When we see the sun going down, we are looking west, because the sun sets in the west. But on Venus, the sun rises in the west and sets in the east! Remember, the reason the sun seems to rise and set is that the planet is rotating. The sun is not actually moving in the sky, the part of the planet you are on is just turning towards and then away from the sun. Since Venus rotates opposite of earth, the sunrise and sunset are opposite as well.

On earth, the sun sets in the west. Because Venus rotates opposite the direction earth rotates, the sun sets in the east on Venus.
Photo by Mark Whitney

This is more important than most people realize. You see, some scientists think the universe began by a big accident in space. The way they describe this accident is that all the planets just formed out of a huge explosion (they call it the "big bang") that sent particles into space. Those particles then eventually formed stars and the planets that orbit around the stars. Well, if this were so, all the planets would have been formed spinning in the same direction. Venus's rotation is strong evidence that this did not happen. Venus spins in the opposite direction of most planets, so it is hard to understand how it could form from the same bits of dust that some people think all things came from. Try to remember that when someone tells you they believe that everything began as a big accident in space.

Believe it or not, a year on Venus is shorter than a day on Venus! This means that Venus orbits around the sun faster than it rotates! You see, it takes Venus only 225 earth days to orbit around the sun, but it takes 243 earth days for Venus to make a complete rotation. If you were on Venus, then, by the time day turned into night and back into day again, *more than a year* would have passed! This is not true for any other planet.

Now I want you to think back to how long a year is on Mercury. Look in your notebook and see if you can find out how many earth days it takes Mercury to make one orbit around the sun. Now compare that to the length of a year on Venus. Which is longer? A year on Venus is longer than a year on Mercury. It turns out that the farther a planet is from the sun, the slower it travels around the sun. Because of this, the farther a planet is from the sun, the longer its year will be. Since Mercury is closest to the sun, its year is very short. Can you guess which planet has the longest year? Think back to what you learned in Lesson 1. Which planet is farthest from the sun? The answer is Pluto. That means Pluto has the longest year. You will learn how long a year is on Pluto later on in the course.

Can you explain in your own words all that you have learned about Venus so far?

Not a Twin

Venus is right next to the earth. It's also about the same size as the earth. Long ago, before anyone knew much about Venus, scientists said Venus was a twin of earth. Many scientists back then thought there might even be dinosaurs or other creatures roaming around on Venus. Today we know that Venus is very different from the earth. No plants or animals could ever live on Venus, because it is too close to the sun and has that terrible atmosphere. This just shows you that looks can be deceiving. Looking at Venus through a telescope made people think that Venus and earth were "twins." Now that scientists have studied Venus "up close" using unmanned spacecraft, we know just how deceiving looks can be!

This is a size comparison of Venus (left) and the earth (right). Venus is colored blue just to emphasize how much it looks like the earth.

Spacecraft to Venus

Do you remember how many spacecraft have visited the planet Mercury? There has been only one. However, 22 spacecraft have visited Venus! All of these spacecraft have been unmanned, but they were equipped with lots of scientific instruments, so we know a lot more about Venus than we do about Mercury! The first spacecraft to visit Venus was Mariner 2, way back in 1962. Do you remember the year your mother was born? You were supposed to learn that in Lesson 2. If you don't remember, ask her again. If she was born before 1962, do the same kind of subtraction that you did in Lesson 2 to find out how old she was back then. If she was born after 1962, she had not even been born when Mariner 2 visited Venus!

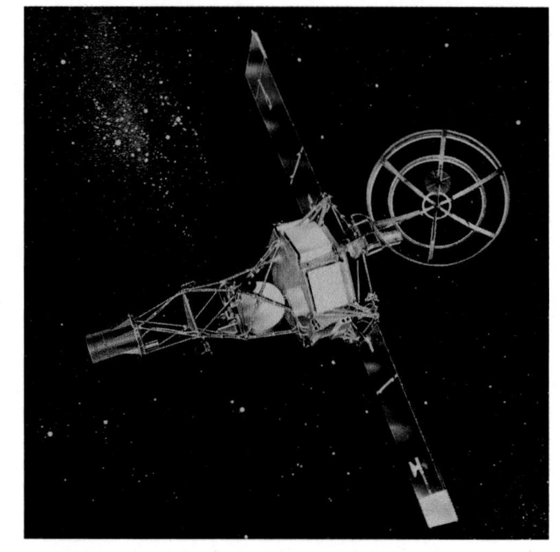

This is an artist's idea of what Mariner 2 looked like as it flew through space to visit Venus.

Now remember, Venus is very hot and is covered with sulfuric acid clouds. Because of this, most of the spacecraft that visited Venus never tried to land on its surface, because the conditions on the planet would just destroy them. Instead, most of the spacecraft simply orbited around the planet, using their scientific instruments to gather as much information as possible. That was actually a very hard thing to do. Why? Well, it is hard to see through the clouds that surround Venus, which means it is very hard to see the surface. Of course, scientists wanted to learn as much as possible about the surface of Venus, so they used **radar** (ray' dar) to find out what Venus looked like below the clouds, because radar can get through the clouds.

What is radar? Well, a radar unit shoots out signals and then waits for those signals to bounce off of something and come back. It measures the time it takes for this to happen, and from that, it figures out how far away that something was. In the case of spacecraft orbiting Venus, the radar unit sent signals down to the planet, and it waited for them to bounce off the surface of the planet and then come back. From that, it determined how far away the surface of the planet was.

What good does that do? Think about sending a radar signal to one spot on the planet, and then think about sending another radar signal to a second spot just to the right of the first spot. Suppose the radar said that the first signal traveled 500 miles and that the second signal traveled 505 miles. What would that tell you? It would tell you that the second spot on the planet is five miles lower than the first spot, because the signal traveled five miles farther before bouncing back. In other words, there is a cliff there that is five miles high! If you use radar carefully like that, you can figure out what the surface of the planet looks like, even though you cannot see it because of the thick clouds. The project at the end of this lesson gives you a little experience working with something that is very much like radar.

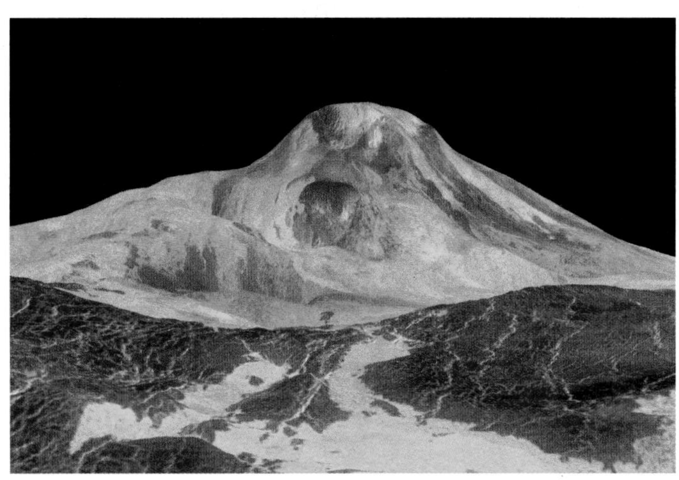

This is an image of a volcano and the valley below it on the surface of Venus. It was made using radar.

Although many of the spacecraft that visited Venus did not land on the surface of the planet, some did. The temperature and conditions of the planet were hard on the spacecraft, and some were destroyed either as they were in the process of landing or once they had landed. Even though these spacecraft were destroyed, scientists did learn things from them. For example, a Soviet spacecraft called "Venera 13" was able to land on the surface, but it stopped working shortly after it landed. Before it stopped working, however, it was able to send two photographs back to earth. One of those photographs is shown above. The picture you see on top is what was taken by

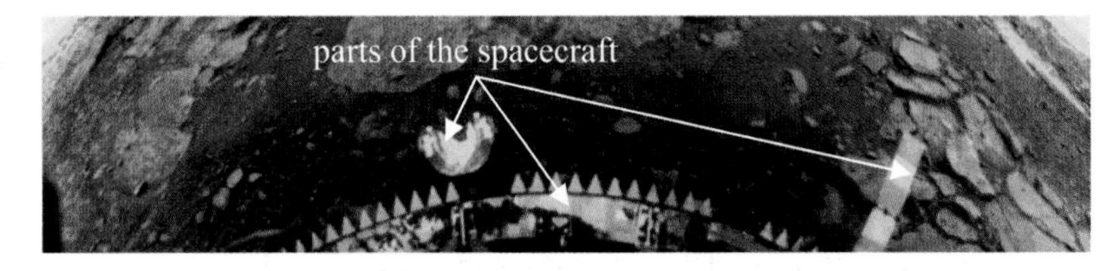

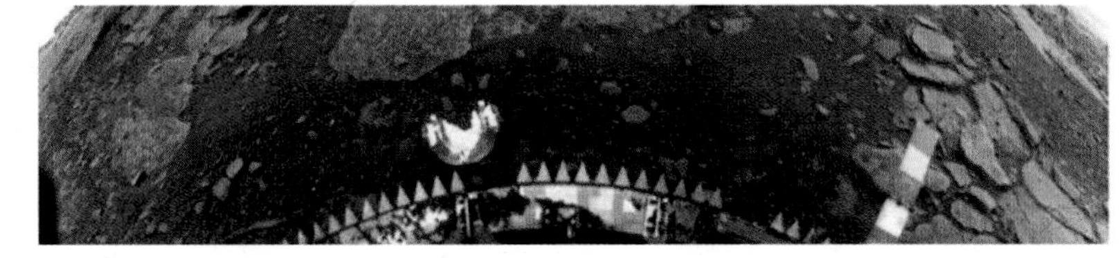

This is two different views of a photo sent back from the surface of Venus.

the spacecraft. The picture on the bottom is the same, but the effects of Venus's atmosphere are removed. In other words, this is what the picture would have looked like if Venus had no atmosphere. The difference between the pictures shows you how Venus's atmosphere "colors" everything on its surface.

The Phases of Venus

If you look at Venus through a telescope, you will see that it appears to change shape from day to day. Galileo discovered this. Do you remember who Galileo was? He was the man who first began studying space with a telescope. He noticed that when he looked at Venus with his telescope, sometimes it looked like a curved sliver, called a **crescent** (kres' unt); sometimes it looked like half of a disk; and sometimes it looked like a full disk. At other times, it could not be seen at all. He called these shapes the **phases** of Venus, and he correctly concluded that this was strong evidence that Venus orbited around the sun.

In this drawing, you can see what Venus looks like to someone on the earth at different spots in its path around the sun. Sometimes it is completely black because it is between the earth and the sun. Remember, the planets do not shine with their own light. They must reflect the light of the sun in order to be seen. If Venus is between the earth and the sun, it reflects light back to the sun, not to the earth. This means Venus looks dark to us, because we do not see any light coming from it. During those times, then, we cannot see Venus in the night sky.

This drawing shows you why Venus has phases when you view it from the earth with a telescope.

When Venus is behind the sun, it looks like a bright disk, because the side facing the sun reflects light back to earth. Because of this, we see an entire side of the planet, and it looks like a disk. When Venus is to one side of the sun or the other, only about half of the planet can reflect light back to the earth, so we see the planet as half of a disk. At other points in its orbit around the sun, only a sliver of Venus can reflect light back to the earth, so we see Venus as a thin crescent. At other points in its orbit, all but a sliver of the planet can reflect light back to the earth, so we see most of Venus, with just a sliver removed. Of course, Venus is so far away that it just looks like a point of light to you and me. However, if you look at Venus through a telescope, you will see these phases quite clearly.

Look at the drawing of the phases of Venus again. Does that look familiar to you? It should. If you have spent much time looking at the moon, you will see that its shape seems to change, much like what is shown in the drawing. That's because the moon goes through phases, just like Venus. We will talk about the phases of the moon in detail later on in this course.

Finding Venus in the Sky

Because Venus is close to the sun, we can only see it for a while before the sun rises and for a while after the sun sets. Of course, since Venus goes through phases, we won't always be able to see it in the night sky. When it is visible, however, Venus is the brightest thing in the night sky, except for the moon. It is often called the "morning star" or the "evening star," because it shines brilliantly after sunset or before sunrise. In fact, this is the reason the Romans named it "Venus." It shined so brilliantly in the sky that they thought it was the most beautiful thing in the sky. As a result, they named it after their goddess of beauty, who was called "Venus."

On a very clear morning or evening, look toward the early rising or setting sun. You might see a bright point of light that looks like a star. That's Venus. Of course, you will know that it's not a star at all. It's a burning hot planet with lava and heat-trapping clouds made of sulfuric acid swirling madly around it. If you need help finding Venus, go to the website I told you about in the introduction to the course. It will tell you where to look for Venus and whether or not Venus is visible when you are looking for it.

What Do You Remember?

What do you remember about Venus? Why did astronomers think Venus was a twin of the earth? What would it feel like on Venus? What is the atmosphere like on Venus? What is special about the rotation of Venus? Have very many spacecraft visited Venus? Since we can't see through the thick clouds over Venus, how do we know what the planet's surface looks like? Why does Venus go through phases? Tell someone all that you learned about this planet. That way, you will be less likely to forget the interesting things you have learned.

Assignment

Illustrate a picture of Venus for this part of your notebook. Write down the interesting things that you have learned about Venus.

Activity

Have you ever seen a comic strip? It is a story told inside little boxes. If you have a newspaper in your house, you can see what a comic strip looks like. Today you are going to make a comic strip called "A Day on Venus."

Think about what story you would like your comic strip to tell. Remember how long the day is on Venus. Remember the temperature on Venus, and think about how you would feel if you were there. Would you have to have a special suit on? Would you want to go home? These are all things that you could include in your comic strip.

Make a sketch on a piece of paper of what you will put in each box before you do the real thing. When you are ready to do the real thing, make boxes like those in a comic. Your comic strip can have as many boxes as you wish, as long as there are at least two. Here are what comic strip boxes look like:

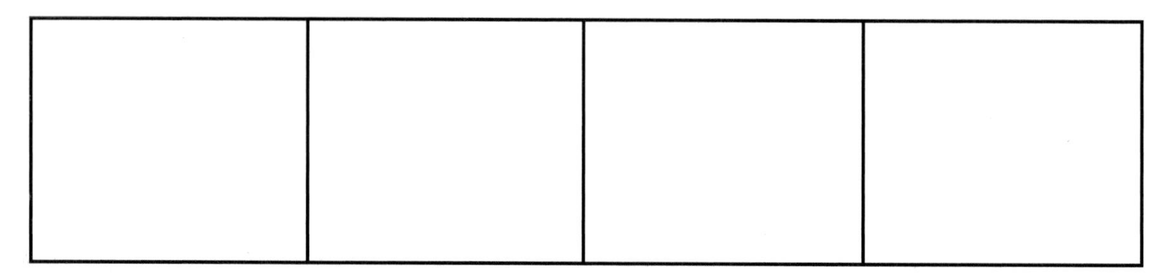

After you have planned out your story, you can begin drawing inside the boxes. If you put people in your story, make "call out" boxes for their words. A "call out" box is a little box coming from the mouth of the person who is talking. It contains the words that the person is saying. The drawing on the right shows you what I mean.

Put your comic strip in your notebook when you are finished.

Project
Learn How Radar is Used

Because a thick layer of clouds covers Venus, spacecraft cannot see the surface. However, as I told you earlier, they can use radar to determine what the surface looks like. In this project, you will use a method that is something like radar to determine what the inside of a box looks like, even though you cannot see into it.

You will need:

- A shoe box or other square box
- A piece of cloth, such as a dish cloth or strong paper towel, that will cover the box and drape down its side. The cloth should be thick enough that you cannot see through it.
- A large rubber band and perhaps some tape
- Newspaper
- Plaster of Paris (when mixed with water it should dry in an hour) or a glue mixture (half flour and half water - this will take a full day to dry)
- Bamboo skewer (or a long, skinny stick with a point)
- Markers, crayons or colored pencils (Markers work best.)
- Ruler or measuring tape
- A parent to help you

1. Ask your parent to crumple newspaper into the bottom of the box. Do not watch her while she does this! You are not supposed to see what is in the box. Ask her to leave a few inches of flat space somewhere in the box, but she should also make some hills and valleys with the newspaper.
2. After your parent has done this, she needs to pour the plaster (or flour/water mixture) over the paper. She then needs to let the whole thing dry.
3. Once the plaster or glue has dried, your parent needs to place the cloth over the box and secure with the rubber band. To make the cloth covering very secure, she might want to tape it to the box. You will be sticking the skewer through the cloth, so it needs to be very secure on the box. Once again, you should not have looked into the box at all. You should have no idea what is inside the box.
4. Look at what you have. You know that there is something in the box, but you can't tell what it is, because the cloth hides it. However, by using the "radar" in this experiment, you will be able to get a good idea of what is in there.
5. Begin by marking a grid on the cloth that is covering the top of your box. Do this by making rows and lines in both directions. The more lines you have, the more accurate your ability to detect what is in the box. Number each box on your grid, beginning with 1. The drawing on the next page shows you what I mean. You can also look at the picture on the next page to see what it looks like in an actual experiment.

1	2	3	4	5	6	7	8	9	10
11	12	13	14	15	16	17	18	19	20
21	22	23	24	25	26	27	28	29	30
31	32	33	34	35	36	37	38	39	40
41	42	43	44	45	46	47	48	49	50
51	52	53	54	55	56	57	58	59	60

6. Copy the grid on the cloth onto a blank sheet of paper. This is where you will record your data.
7. Use the markers to color code your bamboo skewer. Beginning from the bottom, color the first inch black, the next inch blue, the next inch green, the next inch red, the next inch yellow, the next inch orange, and so on, up to the depth of the box.
8. Begin by sticking the bamboo skewer though the first box on the grid. Once it goes through the cloth, push it down gently until you feel it touching something inside the box.
9. Look at the skewer to see the color that is closest to the cloth. Fill in that color on the first box of your paper grid.
10. Continue to do this for each box on the grid.

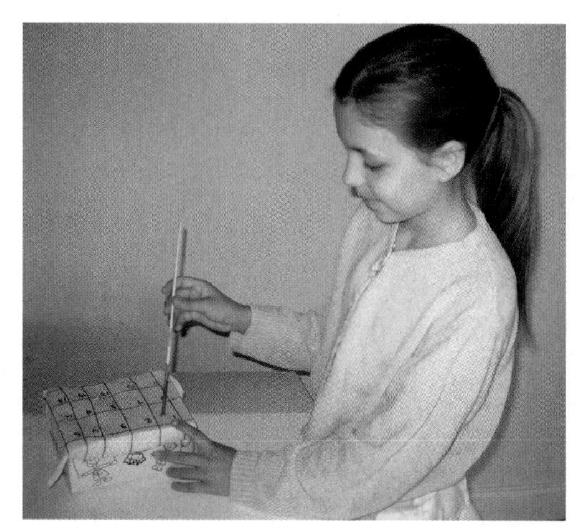

Look at your paper grid. Each box has a color in it. Find a box with black in it. What does that mean? It means that the skewer only went down about an inch or so before it hit what was in the box. What does that tell you about what is in the box at that point? Whatever is there must be pretty tall, because the skewer did not go very far down into the box. Now look for a box that has red or yellow in it. At that point, the skewer went down much farther, so whatever is in the box at that point is not very tall. Now look for a few boxes that all have the same color. What does that tell you? The land must be flat in that area, because the skewer went down the same distance in all of those boxes. If you wanted to land a spacecraft on the surface you just mapped out, you would want to find a flat area like that. Look inside the box to see how the surface in the box matches your paper grid.

Think about what you have done here. Without seeing what was in the box, you were able to determine the hills and valleys that were in the box, and you were able to find a flat spot to land a spacecraft. This is how scientists use radar!

Lesson 5
Earth

Perfect Design by a Perfect Designer

"The heavens are the heavens of the LORD, But the earth He has given to the sons of men."
Psalm 115:16

Now we get to take a closer look at the most fabulous planet in the entire solar system! I know, I know, the earth may seem a bit boring since you see it every day and already know a lot about it. But it's actually the most amazing planet in our whole solar system! God's fingerprints are all over the earth. In other words, when we look at the details that make the earth so perfect for us, it's obvious that only a very wise and wonderful God could have made it. Our planet is so special, so perfectly suited for people, plants, and animals to live, that it would be impossible for it to have happened by accident. In fact, scientists say it would be next to impossible for another planet like ours to exist anywhere in the whole universe! There are many factors that God wove together to create the earth so that it supports life. If one of those things were missing, life could not exist. This means that the earth is almost certainly the only planet in the whole universe that has any life on it.

The Bible tells us that God made the earth. Genesis 1:1 says, "In the beginning God created the heavens and the earth." Remember we talked about how God makes everything for His glory? Well, if the beautiful heavens declare his glory, the earth must be screaming His glory, because it is the most glorious of all.

Perfect Distance

Earth is in the zone! There's a small zone around the sun where life can exist. God placed the earth right in that zone - the most perfect place in the entire universe for us. We're not too far away or too close to the sun. The earth has plenty of water, which is the main ingredient for life to exist. Water is everywhere on earth. There is even water inside of you!

This is a model of the earth in front of a picture of the sun.

Do you know that if we were closer to the sun, the oceans would disappear? They would just dry up! The atmosphere covering earth would also be destroyed, and the harsh rays from the sun would burn us until we died. We could not survive on this planet if it were much closer to the sun.

If we were farther away from the sun, it would cause other terrible problems. The water would all freeze, and we would have frigid, icy weather every single day of the year. Spring would never bring new plants, and the animals would eventually die of starvation. Even a small change in the earth's temperatures would cause worldwide disaster. God placed us exactly where we need to be. He made the earth a perfect place for people to live!

Perfect Mass

Do you have a baseball and a tennis ball around the house? If so, go get them. Hold the tennis ball in one hand and the baseball in the other. These two balls are almost the same size, but as you hold them, you should notice that the baseball is heavier than the tennis ball. Why is that? If they are pretty much the same size, why is one so much heavier than the other? The answer is that there is more **matter** in the baseball than there is in the tennis ball.

What is matter? It is the "stuff" that makes up everything around you. Everything you can touch or smell (including yourself) is made of matter. The more matter that is in something, the heavier it is. Since the baseball is heavier than the tennis ball, the baseball must have more matter in it. Now remember, these two balls are pretty much the same size. So, if the baseball has more matter in it, what does that tell you? It tells you that matter is "packed" into a baseball more tightly than matter is packed into the tennis ball. Think about putting toys into a box. If you put a ball and a squirt gun into the box, it wouldn't be very hard to lift. However, if you took all of your toys and stuffed them into the box, it would be a lot heavier, wouldn't it? Why would the box be heavier? It didn't change size. It just had more things packed into it. In the same way, the baseball has more matter packed into it than the tennis ball, so it is heavier than the tennis ball, even though it is pretty much the same size.

This balance tilts towards the baseball because the baseball has more mass than the tennis ball.

Well, the amount of matter in an object is a pretty important thing to know. Because of that, scientists have come up with an idea called **mass**. Mass is a measure of how much matter is in something. If something has a large mass, it has a lot of matter in it. If something has a small mass, it has only a little matter in it. This tells us that the baseball's mass is greater than the tennis ball's mass, because we already figured out that the baseball has more matter in it.

Why am I telling you about mass? Well, the mass of a planet is important because it determines the amount of gravity the planet has. Do you remember what gravity is? We talked about it in Lesson 1. It is a force that planets use to pull on things. The sun's gravity pulls on the planets, keeping them in their orbits. The earth's gravity pulls on you, keeping you on the ground so that you don't go flying off into space. Even if you try to jump as high as you can, you won't go very high, because the earth's gravity pulls you back down to the ground.

Now think about what would happen if God had made the earth with less mass. If the earth had less mass, it would have less gravity. This would mean it wouldn't pull on us as hard. We would be much lighter. Running a mile would be easy. We could jump onto the roof of our home without any effort. We could jump up into any tree and if we fell, it wouldn't hurt very much.

Apples fall from trees because of gravity. If the earth had more mass, apples would fall more quickly. If it had less mass, they would fall more slowly.

However, it wouldn't be all fun and games for us if the earth had less gravity. Remember, gravity keeps things on the ground. The wind would be able to blow things like cars and people right off the ground and into space! The earth would not be able to pull rain out of the clouds and onto the ground, so it would never rain! That would cause severe droughts and famines over the whole world. Phew! It's a good thing earth doesn't have less gravity. Since the amount of mass determines the amount of gravity, we can also say it's a good thing that earth doesn't have less mass.

It's also a good thing that the earth doesn't have more mass. If it did, the gravitational pull would be too strong. It would be a lot of work just to walk around. We would tire very easily. Going up a flight of stairs would make us pant and need a long rest. Many dangerous chemicals that are harmful to breathe, like methane and ammonia, would just sit here on the earth instead of rising up into the atmosphere and floating into space. Have you ever smelled ammonia? If the earth had more mass and you spilled a bottle of ammonia in your house, it would poison your home because the fumes would stay low to the ground where you are. They would spread throughout your house. Have you ever smelled a freshly painted house? If the earth had more mass, that terrible smell would never go away. Because of the earth's perfect mass, many dangerous chemicals rise up out of the atmosphere and into space, leaving us with clean, safe air to breathe.

Can you explain in your own words all that you have learned about the earth so far?

Perfect Rotation

Another special feature of the earth is its **rotational period**. A rotational period is how long it takes a planet to rotate a complete turn. Since a planet's rotation turns night into day, a planet's rotational period is the length of one full day on that planet. The earth's rotational period is perfectly timed so that when we get tired, it gets dark outside. When we have slept enough, the sun comes up, making it light again. We have a 24-hour day, with about 12 hours of daylight and 12 hours of darkness. That's not the best thing about the earth's perfect rotational period, however!

Do you realize that how quickly or slowly a planet turns also affects the weather? If earth rotated more quickly, the winds would be so strong that there would be tremendous hurricanes everywhere. Survival would be difficult. On the other hand, if the earth rotated too slowly (giving us longer days and longer nights) the temperature shift would be extremely hard on the environment. You see, the longer the nights, the colder they get, and longer the days, the hotter they get. A long rotational period, then, would result in cold, cold nights and hot, hot days. The rotational period that God ordained for the earth keeps the temperature in balance to protect the people, animals, plants, and other creatures that He created for His glory.

This is a satellite image of Hurricane Andrew on August 25, 1992. If earth's rotation were faster, more hurricanes would occur.

The Bible says in Isaiah 45:18, "He is the God who formed the earth and made it, He established it and did not create it a waste place, but formed it to be inhabited." It isn't by random chance or accident that the earth has the perfect rotational period, the perfect mass, and the perfect position in space. All the other planets were made to be empty, but not the earth. God designed the earth to be inhabited. That means it was made for people, animals, plants, and other creatures.

Perfect Atmosphere

Another wonderful thing about earth is the special chemicals that God placed in the atmosphere. We've already talked a little bit about what an atmosphere is. In Latin ***atmos*** means water vapor or ***steam***, and ***sphere*** means ***ball***. If you put those two words together, you have ***ball of steam***. That's why we call the mist and gases surrounding a planet the atmosphere of the planet. On earth, we often call our atmosphere the "air."

Even though you cannot see the atmosphere, it is definitely there! Take a deep breath. Do you realize you just sucked in lots and lots of **oxygen** (ox' uh jen) molecules? Oxygen is the most important thing in our atmosphere. Without oxygen, we would not be able to breathe. Astronauts take huge tanks of oxygen with them when they go out of our atmosphere. Their spacesuits pump this oxygen into their helmets so they can breathe it. If they did not have these oxygen tanks, astronauts would not be able to breathe in space.

This is a photograph of an astronaut in space. Note the spacesuit (which has the oxygen tanks) and the earth below.

Have you ever been inside a **greenhouse**? A greenhouse is a "house" where gardeners grow plants. Young plants are often grown in greenhouses, because they are very fragile and must be protected from hot or cold temperatures. Most greenhouses have a roof that is milky white but at the same time somewhat clear. This allows the sun to shine on the plants, while at the same time, protecting the plants from getting too hot. A greenhouse also keeps the plants from getting too cold. Our atmosphere is like an invisible greenhouse that protects us from very high temperatures and very low temperatures.

This is another reason astronauts must wear spacesuits when they are in outer space. The spacesuits keep them warm. It is freezing in outer space, because once you leave our atmosphere, you leave the comfort and warmth that God has provided us. Without a spacesuit, astronauts would freeze in outer space.

Our atmosphere goes very high up in the sky: about 800 miles. That is very high. It's more than 4 million feet. Airplanes don't even go up that high. Only spaceships and astronauts leave our atmosphere.

None of the other planets have the kind of atmosphere that we need to stay alive. Remember, Mercury doesn't have an atmosphere, and because of that, it freezes at night and burns hot during the day. Venus, on the other hand, has an atmosphere that would kill us if we tried to breathe it. Also, Venus is always extremely hot whether it is day or night, because the heat cannot escape its atmosphere. As you learn about the other planets in our solar system, you will find that none of them has an atmosphere that would keep us alive. That's okay, because the Bible says that the highest heavens belong to God. God did not make any other planet to be inhabited by people. No other planet has the protection and security of an atmosphere the way earth does.

Imagine going outside one day to find that rocks were falling out of the sky. What if this happened every day? We wouldn't want to leave our homes, and if we had to, we would need steel umbrellas to protect ourselves! We would be afraid to look up at the sky, and our houses and buildings would have to be much stronger than they are now. Well, the fact is that rocks of every size - giant boulders, little pebbles and teeny tiny grains of sand - are constantly flying into earth everyday! Thankfully, the atmosphere that God has given earth causes most of these rocks to burn up before they reach the ground. This protects us from falling rocks every single day. Most of these rocks become nothing but dust before they land on the ground. If God had not created our atmosphere just as He did, we would have to worry about getting hit by space rocks.

Have you ever seen a **shooting star**? Some people might call it a "falling star." Well, it turns out that shooting (or falling) stars are not really stars at all! They are actually space rocks that hit the earth's atmosphere and burn up. Occasionally, a piece of one of these space rocks will actually survive the burning process and make it to the surface of the earth. Sometimes, the piece is big, but usually it is small. If a piece does survive and reach the surface of the earth, it is called a **meteorite** (me' tee or eyet). Since oceans cover most of the earth, meteorites usually fall into the sea. It is rare for one to hit land. We will discuss meteorites and space rocks in Lesson 8.

Try to put into your own words the things you have learned about the earth's rotation and atmosphere.

Perfect Tilt

What is the temperature usually like in the summer? What is it like in the winter? It feels warmer in the summer and colder in the winter, doesn't it? In fact, you probably already know that we have four seasons on earth: winter, spring, summer, and fall. Many people don't understand why we have different seasons. They think that maybe it's hot in the summer because we are closer to the sun in the summer. That is not true. Earth is actually a tiny bit further away from the sun when it is summer in the United States! The real reason it's hotter in the summer is that the sun's light is shining more directly on us during the summer. It's the concentration and aim of the sun's light that makes it hot. In the other seasons of the year, the sun's light is not shining directly on us. Because of this, the sun does not warm our portion of the earth as well, so the weather is not as warm.

Why does this happen? Why doesn't the sun's light shine directly on us all of the time? Well, the earth is permanently tilted in one direction. The North Pole is not straight up north, and the South Pole is not straight down south in relation to the sun. Everything is at a tilt, so the parts of earth that get direct sunlight change with each season, as earth revolves in its orbit around the sun. If you have a globe, you will see that it is tilted. When you turn it, the countries close to the bottom of the globe come up toward the middle just a bit, and the ones near the top of the globe come down toward the middle just a bit.

A good globe shows the tilt of the earth as it rotates.

Do this experiment. Get a flashlight. Pretend the flashlight is the sun. Take it into a dark room and lay it on a table. Take a piece of paper and hold it right in front of the flashlight. When you do that, the paper gets direct light. In other words, the light is intense, concentrated in one spot. Now tilt the paper so that the flashlight shines on it at an angle. Notice that when you tilt the paper, the light is not as intense. It still shines on the paper, but the light is more spread out and less concentrated. That is what happens in the seasons. In the winter, the sun is not shining straight down on us. We are tilted away from the intense sun rays. The spread-out rays that hit us give us less heat, which results in cooler temperatures. The direct, concentrated rays that we get in the summer give us more of the sun's energy, resulting in higher temperatures.

Do the same thing with your globe that you did with the paper. Shine the flashlight directly on the globe, and then tilt the flashlight up, so that the rays spread out. There is more intense heat when the sun's rays are straight and direct.

The **equator** (ih' kway tur) is the imaginary line that divides the entire earth in half. The sun shines almost directly on this middle line all the time. The top half of the earth is called the **Northern Hemisphere** (hem' uh sfear), and the bottom half is called the **Southern Hemisphere**. As the earth revolves around the sun, the hemisphere that points towards the sun gets direct sunlight, while the hemisphere that points away from the sun gets less direct sunlight. In June, for example, the Northern Hemisphere tilts toward the sun. As a result, the sun's rays strike the Northern Hemisphere directly, and it is warm there. How do you suppose it feels in the Southern Hemisphere during this time? Since the Northern Hemisphere points towards the sun during this time, the Southern Hemisphere points away from the sun. Because of this, the Southern Hemisphere experiences *winter* in June. In December, on the other hand, the Northern Hemisphere tilts away from the sun. This causes the sun's rays to strike the northern half of earth less directly. This makes it cooler in the Northern

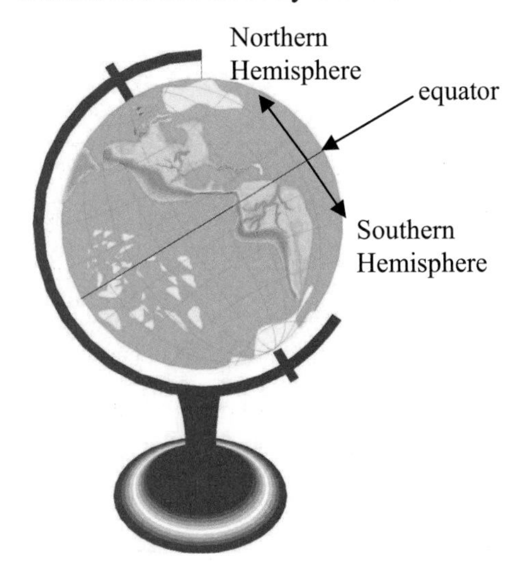

Hemisphere, which is why the Northern Hemisphere experiences winter in December. Of course, at the same time, the Southern Hemisphere points *towards* the sun, so during December, it is *summer* in the Southern Hemisphere!

To better understand this, do an activity with your globe. Take the shade off a lamp and place it in the center of a dim room. The lamp is your sun. Walk your globe around the lamp, as if it is the earth orbiting the sun. As you walk around the lamp, notice the four walls in your room. Each wall will represent a different season. Start so that your globe is between the lamp and the center of one of the walls. We will say that this wall represents winter, so you need to hold your globe so that the Northern Hemisphere is pointing away from the lamp. Notice how the light from the lamp hits the Southern Hemisphere of the globe more directly than it does the Northern Hemisphere. This is why it is summer in the Southern Hemisphere while it is winter in the Northern Hemisphere.

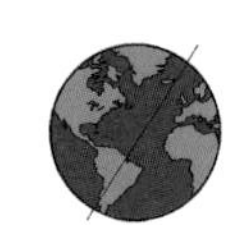

Now walk around to the lamp counterclockwise (that means in the opposite direction from the way the hands on a clock move). Do not change the tilt of the globe. Just watch how the light hits the globe differently as you walk. When you reach the center of the next wall, you will see how light hits the earth when it is spring in the Northern Hemisphere and fall in the Southern Hemisphere. Continue to walk around the lamp counterclockwise, once again making sure that you don't change the tilt of the globe. When you reach the center of the next wall, you will see how light hits the earth when it is summer in the Northern Hemisphere and winter in the Southern Hemisphere. Notice how the lamp's light is hitting the Northern Hemisphere directly now. That's why it is summer in the Northern Hemisphere and winter in the Southern Hemisphere. Continue walking around the lamp until you reach the center of the next wall. At that point, you will see how light hits the earth when it is fall in the Northern Hemisphere and spring in the Southern Hemisphere. The changes that you have seen in how light hits the earth at different points in its orbit around the sun are why we have different seasons on earth.

Remember that the sun shines more directly on the equator as compared to most of the other parts of the earth. This means that countries that are near the equator get the most direct sun rays. What do you think the temperature is like in those countries? They are nice places to go on vacation, because they are usually warm. Find the countries close to the equator on your globe. Places that are close to the equator are warmer than those that are further away. If you live in the southern United States, for example, you are closer to the equator than those who live in the northern United States.

This means you get more direct sunlight, which means all of the seasons are warmer for you. In the southern United States, summer is very hot compared to the northern United States, and winter is reasonably warm compared to the northern United States.

The sun never shines directly on the countries near the top and bottom of the world. Even when the Northern Hemisphere is tilted towards the sun, for example, the North Pole is still too far from the equator to get direct sunlight. What do you think the temperature is like there? Even during the summer, it is still pretty cold (37 to 54 degrees). At the very bottom of earth is an icy continent called Antarctica. Once again, because it never gets direct sunlight, it is also cold. Even when the Southern Hemisphere is pointed towards the sun, temperatures average around 20 degrees. When the Southern Hemisphere is pointed away from the sun, temperatures can get to be 128 degrees *below* zero. During the wintertime (which would be summertime in the Northern Hemisphere), Antarctica is tilted so far away from the sun that it's dark all day and all night for months at a time. Then when it's summer in Antarctica (winter in the Northern Hemisphere), it's tilted so that it is never facing away from the sun. Because of this, it's light all day and all night, for months at a time! Can you believe that somewhere here on earth the sun doesn't go down every night or come up every day? Of course, even though it is light all of the time in Antarctica during the summer, it is not very bright outside. Instead, the sunlight is just a dim glow. Can you explain why the sun is up all day but only looks like a dim glow in Antarctica during the summertime?

This is a drawing of an igloo. Most people think that people who live near the North Pole live in igloos like this. That's not true. Igloos made of snow are usually used as emergency shelters, not permanent homes.

In summer, the days are longer than the nights. This is because the sun is shining down so directly that it takes longer for the earth to rotate enough for us to get out of the sun's light. Also, the sun stays higher in the sky during the summer, because we are tilted towards it. In winter, on the other hand, the days are shorter than the nights. This is because we are tilted away from the sun, so it doesn't take long for the earth to rotate out of the sun's light. In winter, the sun seems low in the sky, because we are tilted away from it.

The earth's tilt is very important to us. If the earth were not tilted, there would be no seasons on earth. The northern United States would always be colder, while the southern United States would always be warmer. That would be a problem, because we grow our food in the summer and allow the ground to rest in the winter. This allows the land to replenish its nutrients, which are like vitamins for the crops. Many beautiful flowers, like tulips, must have a cold winter in order to bloom in the spring. It's part of what makes earth a nice place to live. We need the change of seasons. The cold winter kills the bugs that bite us and ruin our crops. The colder the winter, the fewer bugs we will have in the summer. Places that don't get cold have gigantic bugs!

What is your favorite thing about the cold winter? What is your favorite thing about the warm summer? Without seasons, you could not enjoy those things. Aren't you glad God put the earth on a tilt so that you can see a change in the season?

Do you think you can explain why we have different seasons in your own words? Give it a try right now. Remember to include information about Antarctica and the North Pole in your explanation.

Perfect Land

When I was a kid, my brothers and I had a great big hole in our yard that we dug deeper and deeper every day. We planned to dig until we reached China on the other side of the world. Our hole was so deep we could climb inside it. We didn't understand about the temperatures in the center of earth. We didn't know that if we *could* dig a hole far down into the center of the earth, we would have to go through thousands of miles of molten rock before we ever hit China. The hole that we dug was very deep, but we didn't even get through the *first layer* of the earth.

Our planet has many layers. The top layer is called earth's **crust**, and it contains the oceans, dirt, rocks, and mountains. When you dig a hole in your backyard, you are digging in the earth's crust. Under the crust, there is a layer called the **mantle**. To reach the mantle, you would have to dig a hole in your backyard that is about *20 miles deep*. The mantle is made of hot, semisolid rock, and it is thousands of miles thick. In fact, the mantle is the thickest portion of the earth. The farther down you go, the hotter and hotter the mantle gets. Do you know what magma is? Magma is melted rock. There is a lot of magma in the mantle, because it is so hot. Even below where you are sitting right now, there is hot magma. Don't worry; it is too far down for you to ever see it or feel it. In some places, holes in the crust allow magma to come up to the earth's surface from the mantle. That is where volcanoes have formed.

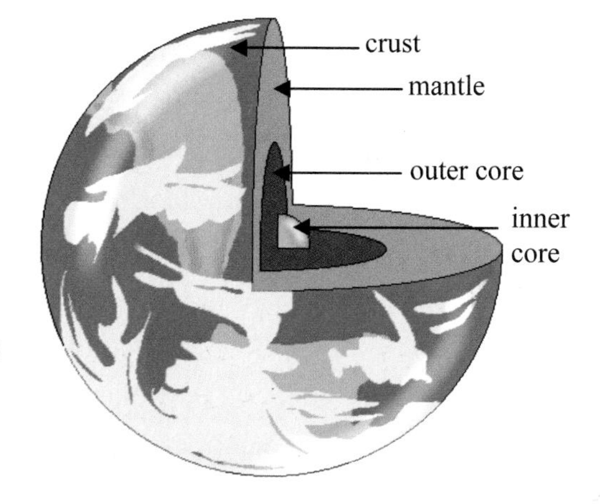

Below the mantle, there is a hot section of melted metals like nickel and iron called the **outer core**. This is an important part of the earth, because the earth's magnetic field is formed here. You will learn more about the earth's magnetic field in the next section of this lesson. The very center of the earth is also made of metals like nickel and iron, but those metals are solid. This solid center is called the **inner core** of the earth.

Perfect Magnetosphere

Perfect magnet-o-what? Remember that "sphere" means "ball." Our **magnetosphere** (mag neet' uh sfear) is like a big ball of magnetic power around earth. It's an amazing, miraculous, mighty feature of our earth. It is simply astounding and reminds us that God thought of everything! He didn't leave any little detail to chance.

Have you ever played with magnets? A magnet attracts certain metal objects. Have you ever noticed that when such an object gets a certain distance from the magnet, it jumps right onto the magnet? That's because the magnet has a **magnetic field** surrounding it, and the object is attracted to the magnet by that magnetic field. Can you believe the entire earth has a magnetic field as well? This magnetic field is produced in the outer core of the earth, and it pulls certain harmful particles away from the earth. These harmful particles make up what scientists call the **solar wind**. Without the magnetic field, we would not survive. God placed this special magnetic field around the earth to protect us from the dangerous particles coming from the sun, keeping us alive and well. Our magnetosphere does an important and life-saving job.

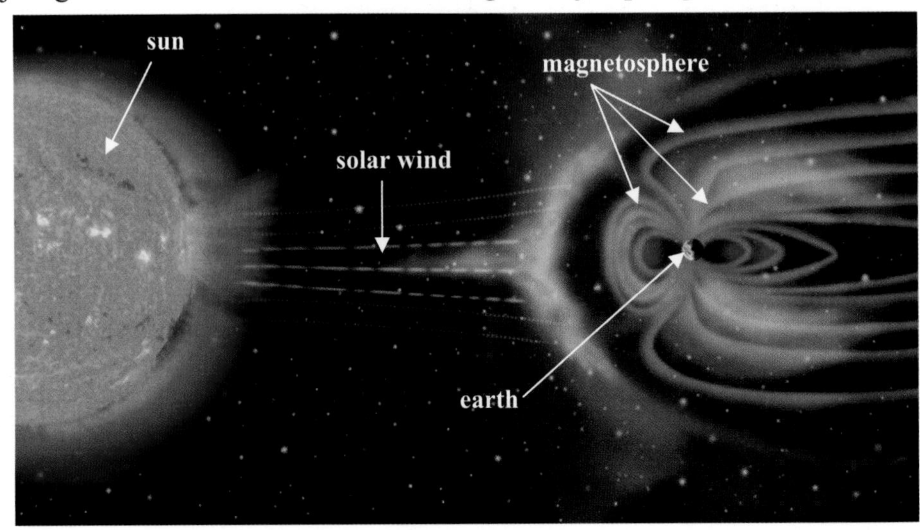

This is a drawing that shows how the magnetosphere protects the earth from dangerous particles that come from the sun.

Although the magnetosphere blocks most of the solar wind coming from the sun, some of the particles that make up the solar wind get trapped in the magnetic field, mostly around the North Pole and the South Pole. As these particles travel in the earth's magnetic field, they start hitting gases in the earth's atmosphere. The energy of the collisions between these particles and the atmosphere's gases produces beautiful colors in the sky. These displays of color are called **auroras** (uh roar' uhs), and they are easiest to see if you live in the northern or southern parts of the globe.

This is a photo of an aurora as seen in the north. Northern auroras are often called the "Northern Lights."

Explain about the earth's magnetic field in your own words.

What Do You Remember?

Can you remember the seven things that make earth the only planet that can support life? Try to explain why those things help us to live on the earth. Why do we have different seasons? What are the four major sections of the earth?

Assignment

Illustrate a picture of the earth for this part of your notebook. Write down the interesting things that you have learned about the earth.

Activity

Today you are going to make an **advertisement**. If you do not know what an advertisement is, ask your parent / teacher to show you one in a newspaper or magazine. Pretend you have been asked to sell the earth to someone who does not know much about it. Make a full-page ad listing all the great things about earth. Be sure to include all the things you have learned. It should be colorful, with different sizes of writing, telling about the different features of the earth. Put this in your notebook.

Project
Make a Compass

A compass is an instrument that always points north. The reason it does this is because the earth has a powerful magnetic field that pulls the needle of the compass toward the North Pole. In this project, you are going to create your own working compass. A compass is important for survival if you ever get lost in the forest. Knowing how to use it will keep you headed in the right direction until you exit the forest. If you were lost and did not have a compass, you might walk in circles as you were trying to find your way home.

You will need:
- A cork
- A permanent marker
- A lid from a yogurt or sour cream container (Try to find one that has a high lip.)
- A sewing needle
- A magnet

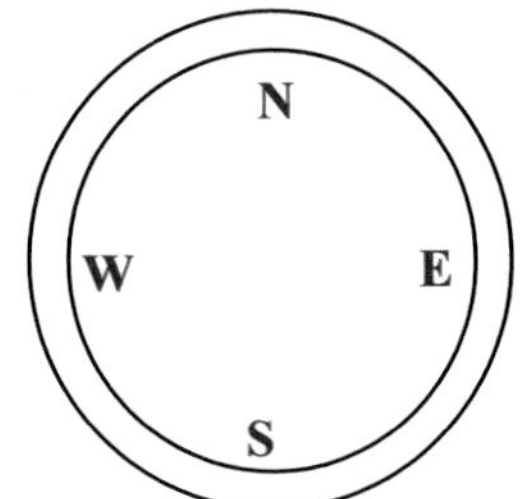

1. Use the marker to label each section of the lid with the first letter of each major direction: north, east, south and west. See the drawing to the right.
2. Run the magnet over the needle a few times, **moving it in the same direction each time it passes over the needle. Do not run the magnet back and forth over the needle**. This action 'magnetizes' the needle a bit. In other words, it makes the needle itself a magnet.
3. Cut off a small, thin sliver of a circle from one end of the cork, and poke the needle through it, from one end of the circle to the other. See the drawing below.

4. Fill the lid with water.
5. Float the cork and needle in the lid so the floating needle lies roughly parallel to the surface of the water. You now have a compass.
6. Place your compass on a still surface and watch what happens.
7. Turn the compass, and notice that the needle continues pointing in the same direction, regardless of how you turn the compass. One end of the needle will always point north, and the other end will always point south.
8. Turn the compass so that the needle is pointing to the "N" that you drew in step 1.

Because you made the needle into a magnet, it is affected by the earth's magnetic field. As a result, one end will always point north, and the other end will always point south.

Lesson 6
The Moon

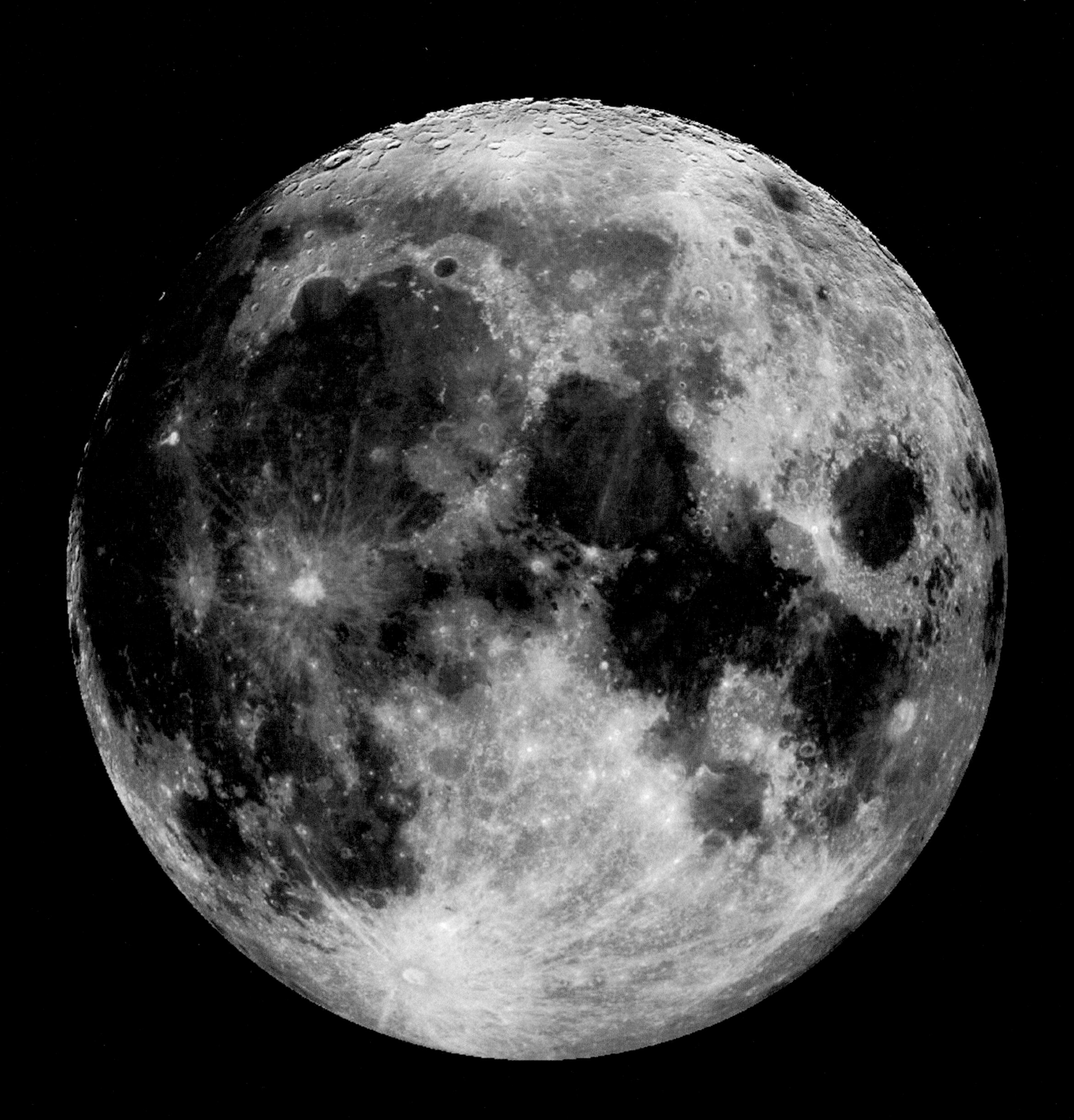

The Moon

Isn't it wonderful to look up at the sky on a clear night? The moon brightens up the night sky. We get our word "month" from the word "moon." Try to say the word "month" with two "o's" in the middle like the word moon: "moonth." Doesn't that sound like month and moon put together? Many years ago, people guessed what day of the month it was by looking at the moon. Another word we use when talking about the moon is **lunar** (loo' nur). It comes from Latin. A lot of scientific words come from Latin. Latin is a good language to learn if you want to be a scientist someday. The Latin word for moon is ***lun***, which is why we often use the word "lunar" to refer to the moon.

Do you remember that the moon is a satellite of the earth? The moon revolves around the earth, like the earth revolves around the sun. We do not call the moon a planet; we call it a satellite. Take a ball, hold it above your head, and move it slowly around your head. The ball orbits your head like the moon orbits the earth. We usually think of satellites as man-made machines that we send up into space to orbit the earth. We use those kinds of satellites to take pictures, collect information, and transmit television and telephone signals. The moon is also a satellite, but it is a *natural* satellite, because it was made by God, not man.

The moon looks like a big light up in the sky. But, actually, the moon is a very dark satellite with no light coming from it at all. "Why is it lit up?" you ask. Well, remember why we can see the planets shining like stars in the night sky. They reflect light that comes from the sun, making it look like they are shining with their own light. Well, the moon does the same thing. The sun is always shining on some part of the moon, so it is always daytime somewhere on the moon. The light we see coming from the moon, then, is actually the light of the sun reflecting off the surface of the moon.

Even though the moon doesn't make its own light, it can be very bright in the night sky. When the moon is big and round, the night is not very dark because the sun's light reflecting off the moon provides some light by which we can see. When the moon is a small sliver, we have darker nights. If you want to play hide-and-seek during the night, it is easier to hide when the moon is either not visible in the sky or when it is just a small sliver.

To understand the fact that the moon's brightness is just the result of how it reflects light onto the earth, try this experiment. Take a compact disc (CD) into a room that has a lamp or a closet with a light bulb nearby. Remove the lampshade from the lamp, and turn on the light. Now move your compact disc in a way that makes the light shine from the compact disc onto your other hand (the one that is not holding the CD). You are reflecting the light from the bulb (which is like the sun) off the CD (which is like the moon) onto your hand (which is like the earth.) The compact disc is not producing light on its own. If you took the compact disc into a dark room, no light would come from it. That is like the moon. The only way the CD can shine light onto your hand is to reflect it from a lamp that is producing its own light. In the same way, the moon reflects the light from the sun onto the earth.

Tell someone all that you have learned about the moon in your own words.

The Moon's Phases

Have you ever noticed that the moon seems to change shape in the sky? Sometimes it's a big round ball; sometimes it's a semi-circle; and sometimes it's a sliver, which is called a crescent. Also, one night during every month, the moon seems to completely disappear.

Well, the moon never really disappears, and it doesn't really change shape. It's always the big round ball that God created it to be. The reason it looks like it has changed shape is because we can only see the part of the moon that reflects light back to the earth. Why do you think only part of the earth and part of the moon are lit up in the picture on the left? It is because the sun's rays are only shining on half of the earth and half of the moon. The bright side of the earth is experiencing daytime, while the dark side of the earth is experiencing nighttime. The people who are experiencing nighttime on earth in this picture would see a full moon, because the entire face of the moon would reflect sunlight onto the earth.

This is a photo of the earth and the moon both reflecting the sun's light.

Do you remember in Lesson 4 when we discussed the fact that if you look through a telescope at Venus, you will see Venus change shape from day to day? Do you remember what we called that? We called it the **phases** of Venus. Well, the reason that the moon seems to change shape in the night sky is that it has phases as well. It's really hard for people to understand how the phases of the moon work. The best way to understand it is to do an experiment.

You will need a lamp and a lightly colored ball (like a baseball or a white Styrofoam ball) on a stick. Put the lamp with its shade removed at one end of a darkened room. Sit at the other end of the room and hold the ball on the stick up in front of you so that it is between your face and the lamp and just slightly above your head. In this exercise, the lamp is the sun; the ball is the moon; and your head is the earth. Now keep your arm straight and slowly spin around in place so that you are constantly looking at the ball and so that the ball is traveling in a circle around where you are sitting. Do this very slowly so you can see how different sections of the ball are lit up. As the ball travels in a circle around where you are sitting, you will see it go through the same phases that the moon goes through. When the ball is between you and the lamp, you see only the dark side of the ball. When the moon is between the earth and the sun like this, it is called a **new moon**. We can't see a new moon because it's totally dark. The night side of the moon is the side facing the earth, and the day side of the moon is reflecting light back to the sun, not at the earth.

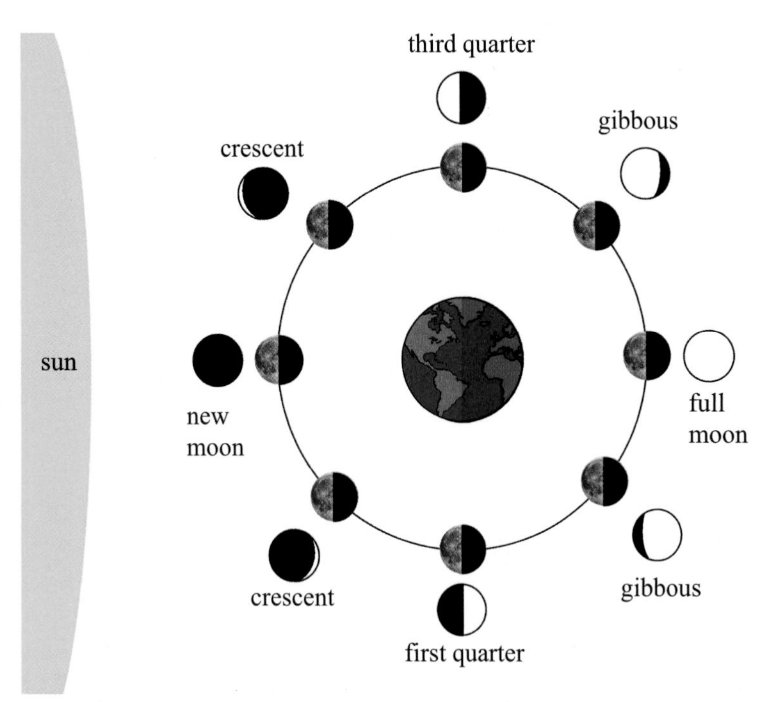

This is a drawing of the moon's phases. The moon is drawn in its orbit. The side facing the sun is bright, while the other side is dark. The drawings beside each image of the moon represent what the moon looks like from earth.

Now look what happens as the ball moves so that it is no longer between the lamp and your face. As it moves, you first see that a small sliver of the ball reflects the lamp's light into your eyes. In the same way, when the moon moves so that it is not directly between the earth and the sun, we begin to see a little sliver of the moon. This little sliver is called the **crescent moon**.

As you continue to spin around, you will get to the point where you can see a semicircle of light reflecting off the ball. When the moon reaches a point like that, it looks like a semicircle in the sky. That is called a **quarter moon**. It's not called a "half moon," even though you see it as a half circle. This is because the word "quarter" refers to the fact that at this point, the moon is ¼ (or a quarter) of the way through its orbit around the earth.

As you continue to spin around, you will see that the portion of the ball that reflects light into your eyes gets larger and larger. When the moon starts getting bigger and bigger after the quarter moon, it's called a **gibbous moon**. As you continue to spin around, you will eventually reach the point where your head is in between the lamp and the ball. Because the ball is slightly above your head, however, your head does not block the light from the lamp, and you see the entire side of the ball shining with the lamp's light. When the earth is between the moon and the sun, we see the entire side shining down on us, so we call it a **full moon**.

As you continue to spin around, you will see the pattern reverse. The bright part of the ball that you see will get smaller and smaller. Eventually, you will see only half a circle, then a sliver, then a dark ball again. The moon goes through exactly the same phases that your ball went through, and it takes about 28 days. That means it takes 28 days for the moon to revolve around the earth. After that, the whole process starts over again. Twenty-eight days is almost a month, which is why the word "month" comes from the word moon.

In the activity at the end of this lesson, you will begin observing the moon every night for 28 days. In your notebook, you will make a calendar and illustrate the phases that you see.

Lunar Eclipse

Every once in a while, as it is making its way around the earth, the moon gets lined up perfectly with the sun and the earth, as shown in the drawing below. When this happens, you could draw a straight line from the center of the sun, through the earth, and to the moon. This causes a **lunar eclipse**. Do you remember what a solar eclipse is? It's when the moon gets in between the sun and the earth, blocking out the sun. Well, a lunar eclipse is when the earth gets right in between the moon and the sun, blocking the sun's light from the moon.

You should never look directly at a *solar* eclipse. However, you can stare right at a lunar eclipse, and it's a beautiful thing to see! When a lunar eclipse occurs, the shadow of the earth crosses over the moon. If you watch a lunar eclipse as it happens, you will see a circular shadow slowly pass over the moon. It will start at one side of the moon and gradually cover more and more of the moon. In a partial lunar eclipse, the shadow covers only part of the moon. In a total lunar eclipse, the shadow covers the entire moon. The portion of the moon covered in the shadow will be a beautiful reddish-copper color. This happens because the sun's light must travel right through the earth's atmosphere before it shines on the moon. Do you remember what happens when the sun's white light travels through the earth's atmosphere? It bounces off of the gases in the atmosphere. Since blue light tends to bounce off these gases more than other colors of light, the blue light does not make it through the atmosphere. As a

This is a photograph of the moon during a total lunar eclipse.

result, it does not reach the moon. Instead, the red, yellow, and orange light from the sun ends up striking the moon, giving it the beautiful color shown in the picture on the previous page.

Do you understand why it is safe to look at a lunar eclipse, even though it is quite dangerous to look at a solar eclipse? It's because the moon does not produce any light. The only light that we see coming from the moon is the sunlight that it reflects. Because of that, the amount of light that comes from the moon is very small compared to the amount of light that comes from the sun. As a result, looking at a lunar eclipse is very safe, and you should try to see one yourself. I have listed some of the upcoming lunar eclipses in the red box. You can learn more about how to see these upcoming lunar eclipses by visiting the course website I told you about in the introduction to the course.

Upcoming Lunar Eclipses

October 28th, 2004
October 17th, 2005
March 14th, 2006
March 3rd, 2007

Lunar Atmosphere

You might hear someone say that a restaurant "has no atmosphere." This means that the restaurant is kind of boring and not very cozy. Sometimes, you'll hear people say, "That place has a lot of atmosphere!" When people say that, they mean the place is warm and friendly. Sayings like these are just expressions, but they do contain a bit of truth. The earth's atmosphere does make the earth a warm and friendly place to live. Remember how our atmosphere is one important thing that protects us and gives us air to breathe? The moon has no atmosphere. So, the moon would not be a very warm or friendly place to live. We would definitely say, "This place has no atmosphere!"

If you went to the moon, everything would be so different from the earth. Without an atmosphere, there would be no protection from the sun's rays. You would need an extra heavy spacesuit so you wouldn't get sunburned. The spacesuit would also carry the oxygen you would need in order to breathe. Because of this, you could only stay outside of your spaceship until your oxygen level got low. You would then have to head back for a refill so that you could explore some more.

This is a photograph of an astronaut walking on the moon. Do you see the very bright glare near the top of the picture? That's from the sun. Notice that the sky is black when the sun is shining, because the moon has no atmosphere.

Do you remember why earth's sky is blue? The blue rays of the sun bounce off the gases in our atmosphere, making it look like blue light is coming from the entire sky. As a result, the sky looks blue. Since the moon has no atmosphere, its sky looks black, even during the day. You should remember from Lesson 3 that Mercury's sky is the same. It looks black even during the day because Mercury also has no atmosphere.

Do you remember how our atmosphere protects us from rocks that fall from outer space? Well, space rocks also fall from outer space onto the moon. Without an atmosphere, all those flying rocks crash right into the moon. Some have been so enormous that they have left huge dents on the surface, which we call craters. The moon is scarred with thousands of these craters.

Interestingly enough, the moon rotates very slowly. It takes just as long for the moon to make one rotation as it does for the moon to make one orbit around the earth. Because of this, the same side of the moon is always facing the earth. Whenever you look up at the moon, then, you always see the same side. Compared to the other side of the moon, the side that faces the earth has very few craters. Instead, it is covered with smooth, flat plains called **maria** (mar' ee uh). They appear as dark patches on the moon's surface. This is a bit confusing to scientists who want to believe that the moon is billions of years old. If the moon were really that old, there should be craters all over its surface, since there is no atmosphere to protect the moon. That means there should be few (if any) flat plains. Of course, if you really want to believe that the moon is old, you can always come up with some reason to explain the flat plains on the moon. Those who want to think that the moon is billions of years old say that the plains have been smoothed out over the years due to changes that have occurred on the moon. If that were true, however, the flat plains should be on both sides of the moon instead of mostly on one side. Of course, a more reasonable explanation is that the moon is not as old as some scientists think.

Walking on the Moon

The moon is the only place in our whole solar system, besides the earth of course, that people have actually visited. The first man who walked on the moon was an American named Neil Armstrong. He went to the moon on a spaceship called Apollo 11 and took the first steps on the moon on July 20, 1969. Do you see the footprint in the picture on the right? That is an astronaut's footprint in the dirt on the moon. Even though this footprint was left on the moon quite some time ago, it is most likely still there. Why? Well, if you leave footprints in the dirt in your backyard, wind will eventually blow the dirt around, filling in your footprints. That will make your footprints disappear. Rain might also smooth out the dirt, once again destroying your footprints. On the moon, however, there is no wind or rain. Because of that, the footprints do not get filled in. If people visit the moon again sometime in the future, they will probably see the same footprints left by astronauts way back in 1969.

This is a photograph of an astronaut's footprint on the moon.

Many spaceships have gone to the moon, but only 12 men have ever walked on the moon. Some of these men not only walked on the moon, but they actually rode a special car that they took with them. It was called the **lunar rover** (see the picture on the next page), and it allowed them to

This is a photograph of an astronaut riding the lunar rover on the moon.

explore more of the moon than they otherwise would have been able to explore. Some people say that we have never actually been to the moon. They say that the pictures we have of men on the moon have all been faked, and that the samples of moon dust and moon rocks are not from the moon at all. There are many arguments which show that these people wrong. If you would like to read more about the people who think we did not visit the moon and the reasons they are wrong, you might visit the course I told you about in the introduction.

Do you think you will ever walk on the moon? If you become an astronaut, you might walk on the moon, or even a planet like Venus or Mars!

The Moon's Gravity

A spacesuit seems like it would be very heavy doesn't it? Interestingly, your spacesuit would not feel very heavy on the moon. That's because there is a lot less gravity on the moon compared to the gravity on earth. Why? Do you remember what determines a planet's gravity? It is the mass of the planet. Well, the mass of the moon is a *lot* smaller than the mass of the earth. As a result, the moon's gravity is weaker than the earth's gravity.

Even though the moon's gravity is weaker than the earth's gravity, it is still there. Because of that, if you were to drop a ball on the moon, it would fall to the ground. However, it would not fall nearly as quickly as it would on earth. If you jumped up as high as you could on the moon, you would still end up falling back to the ground. However, you would go up a lot higher than you could here on earth, and you would not fall back to the ground nearly as quickly as you would on earth. You could jump up and do a somersault in midair and not hurt yourself at all, even if you didn't land on your feet. That's because the moon's gravity would not pull you down as hard as the earth's gravity does.

Because the moon's gravity is so weak, you would not weigh very much on the moon. Look at the chart below to see if you can figure out how much you would weigh on the moon.

Your Weight on Earth	Your Weight on the Moon	Your Weight on Earth	Your Weight on the Moon
20 pounds	3 pounds	70 pounds	Over 11 pounds
30 pounds	5 pounds	80 pounds	Over 13 pounds
40 pounds	Over 6 pounds	90 pounds	15 pounds
50 pounds	8 pounds	100 pounds	Over 16 pounds
60 pounds	10 pounds	150 pounds	25 pounds

If you have ever been to the beach, you might remember walking a long way just to get to the water. If you visited that same beach at another time, you might have noticed that you did not have to walk nearly as far to get to the water. In other words, when you stand on the beach, sometimes the water is near you, and sometimes the water is far away. This is because the ocean has **tides**. When it does not take very long for you to walk down to the water, we say that the ocean is at **high tide**. When it takes a long time for you to walk down to the water, we say that the ocean is at **low tide**.

Believe it or not, the moon's gravity is what causes the tides. It pulls on the earth's oceans, making them bulge outwards toward the moon. Do you know what "bulge" means? If you put your stuffed animals in your shirt, it would bulge out. To "bulge" is to be "pushed outward." The moon pulls on the oceans, and the oceans bulge towards the moon. If the moon makes the ocean bulge towards the shore where you are standing, you see the high tide of the ocean. If the moon makes the ocean bulge towards the shore on another part of the world, you see the low tide of the ocean.

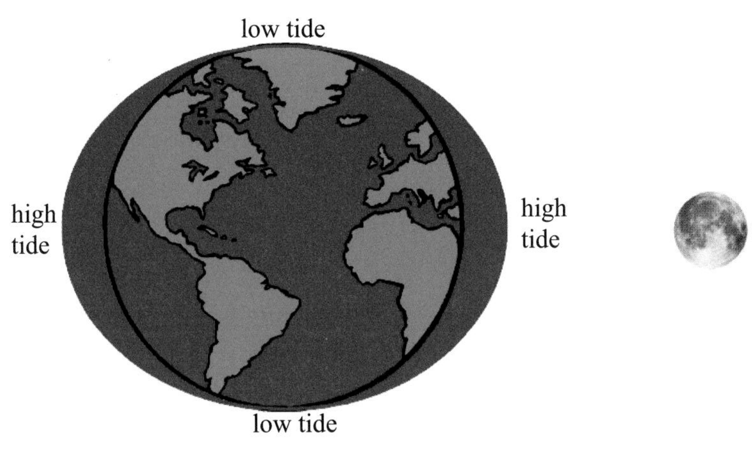

The moon's gravity pulls on the oceans, causing them to bulge out towards the moon. This makes parts of the earth experience high tide, while other parts of the earth experience low tide.

If you have visited the beach, you have probably seen that the ocean has waves. Although the tides can cause certain types of waves, the waves that we see on the shore are not caused by the tides. They are usually caused by wind. Please don't confuse the tides with waves. Tides determine how close the ocean water is to you when you walk onto a beach. At high tide, the water is much closer to you than at low tide. Waves cause the ocean water to heave back and forth. They do not really affect how close the ocean is to you.

If the moon were no longer in the sky, our oceans would not have tides. God gave us an unusually large moon that has enough gravity to pull on the ocean's water. He did this because when the tide comes in and goes out, it cleanses the shoreline. The motion caused by the tides also refreshes the water. Still water can get pretty stagnant, and the tides keep the ocean from getting that way. God planned it so that the earth would have an extra big moon to help keep the oceans fresh and clean.

What Do You Remember?

What is the atmosphere like on the moon? What is the color of the moon's sky during the moon's daytime? Can you explain why the moon has phases? What is a lunar eclipse? Why are the astronaut's footprints probably still on the moon? How does the moon affect the ocean? How is that helpful to the earth?

Assignment

Make an illustration for the moon portion of your notebook. Write down what you have learned about the moon.

Activity

Older Students: Make a calendar of this month to put in your notebook. Ask your parent / teacher to show you what your calendar should look like this month. Every night (and sometimes during the day), look at the moon outside and draw its shape on your calendar. Write down the name of the phase. You can refer to the illustration on page 68 to determine what the shape is called. Keep this in your notebook.

Younger Students: Make a calendar page for your child. You can use a ready-made page from a calendar if you wish, but be certain it fits in the notebook. Have your child draw the shape of the moon on each day of the calendar for a month. Discuss with your child what each phase you see is called. Keep this in the notebook.

PLEASE NOTE: Because of the earth's rotation, your location on the earth faces the moon during specific times. As a result, the moon "rises" and "sets," just like the sun. Sometimes, the moon cannot be seen after sunset, because it sets before the sun. During those times, you can see it before sunrise, because it rises before the sun. You can also see it during the day. The course website discussed in the introduction has a link to a site that will allow you to find out the times that the moon will rise and set in your area. Use this resource to plan when to look for the moon. Try to find the moon in the daytime during this activity as well as at night, so that your students will understand that the moon rises and sets, just like the sun.

Project
Make a Telescope

This is a great exercise to learn about telescopes and how they work. It is fun to create a working telescope that you can use to see the moon (or any other object) better. However, if you want a telescope that will really be able to see far into the heavens, you will need to purchase one from a store. You can build a powerful telescope yourself, but it requires better equipment. You can check out books from the library or find websites that tell you how to make a powerful telescope. This activity, however, will give you an introduction to how telescopes are made.

You will need:

- Two magnifying glasses (One should be stronger than the other. Reading glasses will work also. It works best if you can actually remove the lenses from the magnifying glass holder. This is not necessary; it just makes steps 9 and 10 easier.)
- Cardboard tube (Mailing tubes, paper towel rolls, and wrapping paper rolls all work well. Depending on the magnifications of your magnifying glasses, you might need a long tube.)
- Strong tape (Masking tape and duct tape work best, because they keep light from coming in the sides of the telescope.)
- Scissors (For thick cardboard tubes, you might need your parent / teacher to use a knife.)
- Tape measure
- Paper with writing or an image on it
- Someone to help you

Instructions

1. Place the paper on a table or the floor.
2. Hold the stronger magnifying glass between you and the paper, a good distance away from the paper.
3. Look at the paper through the lens. The image will look blurry.
4. Place the weaker magnifying glass between your eye and the first magnifying glass.
5. Move the weaker magnifying glass forward or backward until the print or image comes into sharp focus. The print or image should appear larger and upside down.
6. Keep the magnifying glasses in place while you have someone measure the distance between the two magnifying glasses. Write down the distance that your helper measures, because that is the distance you will place your two lenses apart from one another within your tube.

7. Place an "X" about 1 inch from the end of the cardboard tube. This is where your first magnifying glass (the one with greater magnification) will be placed.
8. Place another "X" behind the "X" you just made. The distance between these "X's" needs to be the same distance that was measured in step 6. This "X" marks where your weaker magnifying glass will be placed.
9. Cut slots in the cardboard tube where the "X's" are. Do not cut all the way through the tube! Just cut a slot that is large enough so that the magnifying glasses can be inserted into the slots.
10. Place the two magnifying lenses in their slots (stronger one in front, weaker one behind).
11. Tape them in place.
12. Look through the smaller magnification lens at something (**not the sun**) outside. If things are not in focus, you may have to move a lens.
13. Please note that just like the writing on the paper, the images you see through your telescope will be upside down. Despite the fact that they are upside down, the images should appear to be closer than they actually are, because your telescope is magnifying them for you.
14. Try looking at the moon with your telescope. Can you see any details that you were not able to see with just your eyes?

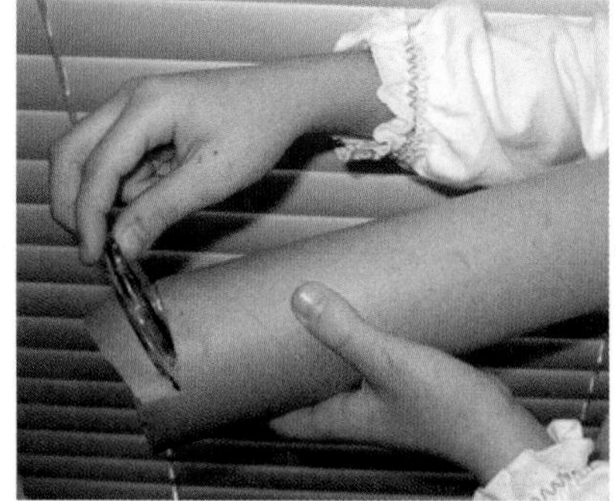

NOTE: NEVER <u>EVER</u> LOOK AT THE SUN WITH THIS (OR ANY OTHER) TELESCOPE. It will severely hurt your eyes! You could even go blind! You must have special equipment if you want to look at the sun through a telescope.

How does this work?

Your telescope works because the magnifying glasses bend the light that comes through them. The first magnifying glass (the one farthest from your eye) gathers the light from the object you are looking at and flips it upside down. The second magnifying glass (the one closest to your eye) takes the image and makes it bigger (magnifying it) for your eye. If you ever want to look at a far away object but do not have a telescope, you can just use two magnifying glasses to bring objects far away into better view.

Lesson 7
Mars

Moving to Mars

Have you ever moved to a new home? It's difficult to get used to a new neighborhood and city. What if one day your parents tell you to pack up your stuff because you are moving to the fourth planet from the sun…Mars? It sounds farfetched, but that is what many scientists hope will happen one day. Because Mars is similar to the earth in some ways, scientists have dreamed about building a community on Mars where people can live.

Although this might sound like a neat idea, there are a lot of problems with it. God did not give Mars all the incredible features that make the earth such a wonderful place to live. Because of this, you would have to take things from the earth that you need for survival. These things would include water, air, heat, food, shelter, and clothing. In fact, you would have to bring just about everything, because there is very little on Mars that would help you to survive there! You would have to live in a special enclosed habitat that scientists have designed, called an artificial **ecosystem** (ee' koh sis tem). This habitat seals in the oxygen, allows the sun's light to come in, and maintains enough water to drink as well as grow plants. What do you think such an ecosystem might look like? This drawing shows one that NASA designed. The living space is underground.

This is an artist's idea of what a man-made ecosystem on Mars might look like.

It will be many years before an ecosystem can be built on Mars, because no one has ever even been to Mars. NASA hopes to send people to Mars at some point, but it costs a great deal of money to send men to other planets. So far, only unmanned spacecraft have been to Mars. Some of these spacecraft have even landed on the surface of Mars and studied many things about the planet. As a result, we know a lot about Mars.

Mars has always been called "the red planet" because it looks red through a telescope, and even to the naked eye. When I say "naked eye," I mean that you are just using your eyes (no telescope, binoculars, or other equipment) to look at it. Robots that have landed on Mars have studied its red dirt, and they have found that it has an element called iron in it. Iron rusts, and that is what makes Mars look red! Mars is, in fact, a rusty planet.

The robots sent to Mars don't look like men walking around with stiff arms and legs, as you might imagine. They are actually little cars with wheels and are similar to toy remote control cars. They are called Mars rovers. These rovers are always bumping into rocks and cannot climb up the mountains of Mars, so engineers are trying to develop other kinds of robots. Some engineers are even working on a robot that looks like a spider. With legs rather than wheels, a spider-like robot might be able to get around better than the rovers.

This is a picture of the Sojourner rover on Mars.

Mars is a terrestrial planet, like earth. It has landforms like mountains, valleys, and volcanoes. One Martian (mar' shin) volcano called **Olympus** (oh lim' pus) **Mons** (mahns) is the largest volcano in our whole solar system. We have never seen a Martian volcano erupt, but there is evidence that some of the volcanoes are active, which means that they *could* erupt. Scientists are not even sure what would come out of the volcanoes if they did erupt. If an ecosystem is ever built on Mars, it should be on level ground that is far from a volcano, don't you think?

This is a picture of a rover on Mars. Notice how red and rocky the landscape is. Notice also that the rover is bumping against a rock.

What do you remember about Mars so far? Explain in your own words the things you have learned.

Martian Gravity

Mars is bigger than our moon, but smaller than the earth. It's small for a planet, and its mass is significantly less than the mass of earth. Because its mass is small, the gravity on Mars is not very strong. What do you think it would be like to live with less gravity? What kinds of things would you have to be careful about if you lived on Mars? With less gravity, a baseball could probably go a lot further when you threw it. I guess you wouldn't want to throw a baseball very

hard, or it might crash through the glass of the artificial ecosystem, letting all of the oxygen out!

How much would you weigh on Mars? Check the chart below. Moving to Mars might not be so bad if you weighed less. It wouldn't hurt as bad if you fell off your bike. It would also be easier to get around if you weighed less. You wouldn't get nearly as tired at the end of the day. Maybe moving to Mars isn't such a bad idea after all!

Your Weight on Earth	Your Weight on Mars	Your Weight on Earth	Your Weight on Mars
20 pounds	Over 7 pounds	70 pounds	Over 26 pounds
30 pounds	11 pounds	80 pounds	30 pounds
40 pounds	15 pounds	90 pounds	34 pounds
50 pounds	19 pounds	100 pounds	38 pounds
60 pounds	Over 22 pounds	150 pounds	Over 56 pounds

Martian Atmosphere

Mars has a very thin atmosphere with no oxygen. That is why we would need to take our own air if we went to Mars. Flying through the atmosphere on Mars are millions of little dust particles, just like in our atmosphere. These dust particles reflect the yellow, red, and orange light from the sun. Guess what color the sky is on Mars. It's butterscotch, which is a yellow/brown color.

This is a photograph of Mars taken by the Spirit rover. Notice the sky color.

Of course, just like on earth, the sky does change colors when the sun rises and sets. The dust particles in the Martian atmosphere make the Martian sunset more of a blue color, as shown in this photograph, which was taken by the Pathfinder rover when it visited Mars.

A Martian Sunset

Do you remember that our atmosphere keeps meteors from hitting our planet? Mars' atmosphere is too thin to offer it much protection from flying space rocks. The bottom half of Mars is virtually covered with craters. The Northern Hemisphere of Mars is flat land with very few craters. If we send a spacecraft to Mars, it will probably be easier to find a good landing spot in the Northern Hemisphere.

All of the craters on Mars make it obvious that, at one time, Mars was being pelted with giant space rocks. When this happened, the force of the collisions probably sent many pieces of Mars flying out of its atmosphere and into space. Some of those pieces of Mars actually landed on the earth! We know this because we have actually found rocks on earth that have chemicals in them which are very similar to the chemicals found in Martian rocks. These rocks also look like they have fallen at high speeds through our atmosphere, so it is reasonable to assume that they actually came from Mars.

Can you tell why this large Maritan crater is called the "Happy Face Crater?"

It is likely that there are also pieces of the earth on Mars. If a giant meteorite hit the earth, the force would be strong enough to send pieces of the earth up into our atmosphere and out into space. Since there are also craters on the earth, we know this has happened before. Now remember, there are not many craters on earth because earth's atmosphere protects us from most space rocks. However, some space rocks are so large that the atmosphere cannot destroy them, and they land on the earth, making a huge crater. Creation scientists believe that if anyone ever finds signs of life on Mars, it will not be Martian life they find, but earth life that made it to Mars! After all, if a piece of earth left our atmosphere, it would take with it many cells and bacteria, which are living things. If we do find life on Mars, then, it will most likely be life that traveled to Mars on space rocks.

This is a photograph of a crater in the Arizona desert.

Moons

If you were standing on Mars at night, you would see not one moon up in the sky, but two! Mars is the only planet in our solar system that has two moons. Astronomers have given them names: **Phobos** (foh' bohs) and **Deimos** (dee' mohs). They are two tiny potato shaped moons. They don't look or act like anything God would have designed to be a moon. Astronomers believe that maybe they were rocks floating around in outer space that got caught in Mars' gravitational pull. This is

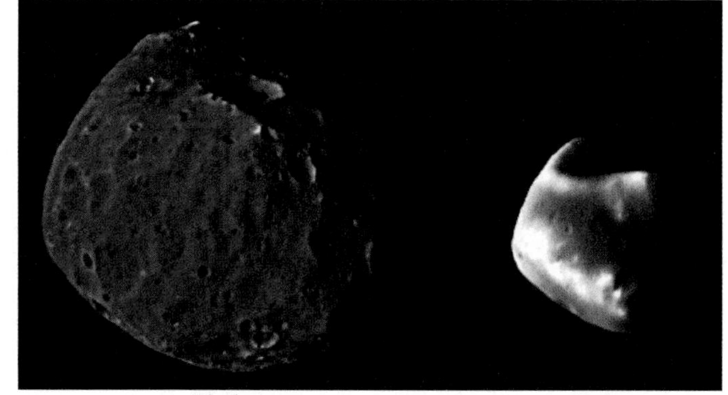

Phobos Deimos

probably true since they don't orbit as a moon would. Unlike normal moons, Phobos an Deimos are getting closer and closer to Mars every day. At some point, they might actually end up being pulled right into Mars. At the speed they are moving towards Mars, however, it would take a long time (about a million years) for this to happen.

Do you remember what a solar eclipse is? It's when the moon blocks out the sun. Well, on Mars, the moons cannot completely block out the sun, because they are too small. However, they can partially block out the sun. While it was on Mars, the Opportunity rover took these photographs of the moon Phobos crossing in front of the sun. You can see that the moon is too small to block out the entire sun, but it does block some of it.

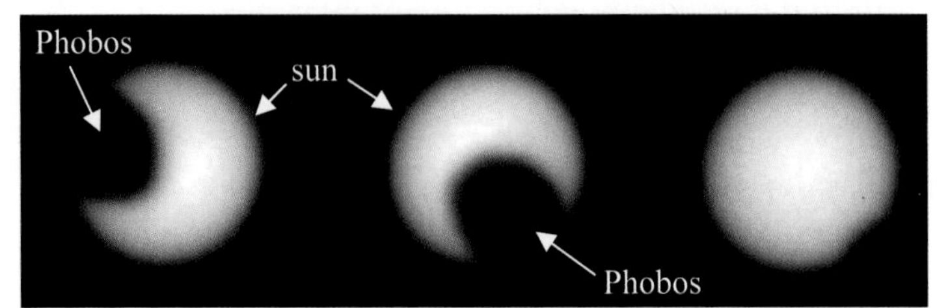

This is a photo of Phobos crossing between the sun and Mars.

If people moved to Mars, would the nights be as bright with two tiny moons as they are with one big moon? What would the night sky look like with a big potato shining down on you?

Martian Orbit

It takes almost two years for Mars to travel all the way around the sun. One year on Mars is 687 earth days. For those who move to Mars, will they count their birthday by earth days or Martian days? What about babies born on Mars? Will there be a different calendar for Martian colonists? If they decided to count time by Martian years, on your 10th birthday, you would really be 20 earth years old. If you lived on Mars, you could drive a car when you were only 8 Martian years old!

If people asked you what nationality you were, would you say, "I'm a Martian," even if your parents are Americans? Would the rest of us have to say we are "earthlings?"

Martian Rotation

If you moved to Mars, possibly the only thing that would not be a huge adjustment for you would be the number of hours in a day. Mars rotates in 25 hours. Earth rotates in 24 hours. So, a day on Mars would be almost the same amount of time as a day on earth. That would really make the whole experience of moving to Mars a little better. It would just be too hard to move to a planet that had days and days of sunlight followed by days and days of darkness. Since Mars has days that are similar to earth, Mars would be an easier planet to live on than Mercury (where the day is 59 earth days long) or Venus (where the day is 243 earth days long).

Mars rotates on a tilt, just as the earth does, so it has four different seasons; however, those seasons aren't much like those we enjoy on earth. Winter on Mars is freezing cold; spring is freezing cold; summer is freezing cold; and fall is freezing cold. That's because it is *always* freezing on Mars! It's just too far away from the sun to get much energy from the sun's light, and its atmosphere is just too thin to keep the heat inside. The average temperature on Mars is usually about 81 degrees below zero, but in the summer, the temperature can get up to about 32 degrees or so. Mars's weather is a lot like the weather in Antarctica (you know…the place where nobody permanently lives because it is far too cold).

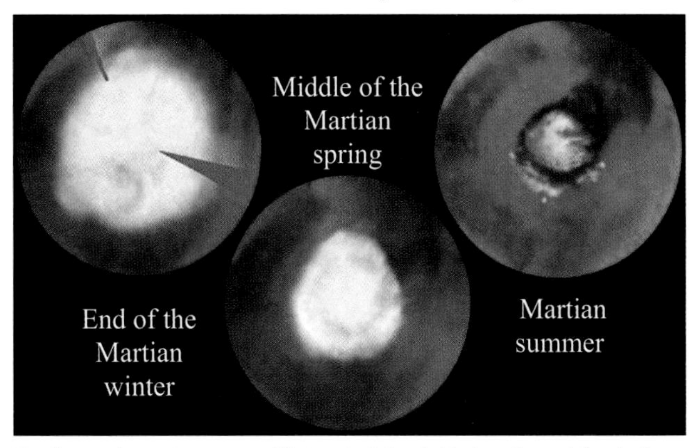

In these three images of Mars's North Pole, the white is ice. Most of the ice is not water ice. It is dry ice (solid carbon dioxide). The fact that the amount of dry ice changes demonstrates that Mars has seasons.

Have you ever seen **dry ice**? Dry ice looks a lot like regular ice, but it is much colder. It is so cold that you would hurt your hands if you tried to hold onto it for very long. Dry ice is actually frozen carbon (kar' bun) dioxide (dye ox eyed'), a gas that you and I breathe out every day. Carbon dioxide has to get very cold to freeze, so when it is frozen, it is much colder than normal ice that you make from water. Well, it is so cold on Mars that carbon dioxide freezes in the winter, making a lot of dry ice.

Scientists think that all of the water on Mars is frozen. Some of it is frozen in **polar icecaps**, which are big slabs of ice at the North and South Poles of the planet. If you look at the picture at the top of this page, the ice that is left on the North Pole during the Martian summer is mostly frozen water. The summer gets warm enough to get rid of the dry ice, but it does not get warm enough to melt the ice that is made of water. Scientists think that the rest of the water on Mars is permanently frozen in the ground. Because the water is permanently frozen in the ground, they call it **permafrost** (pur' muh frost). That sounds like "permanent frost," doesn't it? Well, that's exactly what scientists think it is!

This is a photo of ice on a Martian landscape.

Mars is freezing, but it's closer to the earth's temperature than any other planet in the solar system. This is the reason Mars is the only planet that we could consider sending humans to. It is also the reason that some scientists dream of building a habitat or artificial ecosystem that will protect a community where people can live, work, and grow food. Inside the artificial ecosystem, the temperature would be kept warm. Would you like to live on Mars for a few years? What would you miss about the earth? Would you miss the blue sky? Would you miss the warm breezy summer days? Would you miss swimming in the ocean? If you were to build a protected community on Mars, would you build an imitation ocean for people to swim in? Would you build it around one of Mars' mountains, planting trees and grass on the mountain and providing fake snow in the winter, so people could go mountain climbing and skiing?

Liquid Water on Mars?

Although we know that there is a lot of frozen water on Mars, we do not know whether there is any liquid water on the planet. On earth, there is both liquid water (in lakes, oceans, and streams) and frozen water (in snow and ice). So far, scientists have been able to find frozen water on Mars, but they have not been able to find liquid water there. Right now, as I am writing this book, there are two rovers that are on the surface of Mars. Their names are Spirit and Opportunity. Both of them are exploring the planet, helping scientists learn a lot of things. One thing they are trying to determine is whether or not there is or ever was liquid water on Mars.

What's the big deal? Why worry about whether or not there is liquid water on Mars? Well, liquid water is essential for life, and some people want to believe that there was (and perhaps still is) life on Mars. If they cannot find any liquid water on Mars, or if they cannot find evidence that there *used to be* liquid water on Mars, it will be very hard for them to continue to believe that life once existed on Mars. Of course, if they *do* find liquid water on Mars (or at least evidence that it was there at one time), that does not mean life did exist on Mars. It only means that it would have been *possible* for some form of life to have existed on Mars. Of course, if liquid water did exist on Mars at one time, that means Mars must have been a lot warmer sometime in its past.

How can we tell if liquid water used to exist on Mars? Well, there are some rock formations that are usually made in the presence of water. Also, there are certain chemicals that tend to form in liquid water. Spirit and Opportunity have found both of these things on Mars. The picture on the right is of a rock formation that the Opportunity rover found on Mars. On earth, you see this kind of texture in rock if the rock has been soaking in salty water. The Spirit rover also found traces of a mineral called "jarosite." On earth, this mineral forms in liquid water. So, there is some evidence that liquid water did, at one time, exist on Mars. As far as we know, however, there is no liquid water there now.

This is a photo of rocks on Mars. The texture indicates that the rocks might have been in liquid water.

Once again, it is important to realize that even if liquid water existed on Mars at one time, that does not mean that there was life on Mars. Liquid water is necessary for life, but so are many, many other things. You learned in Lesson 5 that the earth has been perfectly designed for life. Mars does not have most of the design features that the earth has, so it is hard to understand how life could have existed on Mars, even if there was liquid water there. You also need to remember that since space rocks have hit the earth and thrown parts of earth into space, it is possible that there are rocks from the earth on Mars. If that is the case, there may, indeed, be some kind of life on Mars (like bacteria), but that life might have actually come from earth.

Finding Mars in the Sky

Mars is fairly easy to locate in the night sky. It is brighter than most stars, and it has a distinctly orange/red color. In fact, that's how it got its name. The Romans saw that Mars was red and were reminded of blood. As a result, they named it after their god of war, Mars.

Mars moves around quite a bit in the sky, so if you want to find out where it is at any given time, you should go to the website I told you about in the introduction to this course. On that website, you will find links that will take you to websites that will help you find Mars.

What Do You Remember?

What do you remember about Mars? What makes Mars look red? What is the atmosphere like on Mars? What is the surface like on Mars? What is the name of the biggest volcano in our solar system? What do you remember about the moons of Mars? Can you remember how long it takes Mars to revolve and rotate? What is the weather like on Mars? Why do some astronomers think Mars would be a good place to visit and perhaps live?

Assignment

Illustrate a picture for the Mars chapter of your notebook. You can put anything you desire in the illustration. Be certain to label your illustration so others will know what you have drawn. Write down all that you know about Mars and place that in your notebook as well.

Activity

Older Students: Imagine and design a community where people could live on Mars. What would you need to build an ecosystem to protect people and provide the right temperatures to grow food? Illustrate a picture of it. You can use the Internet or encyclopedias to see what ecosystems look like. There is an ecosystem called Biosphere 2 that is located in Arizona. What would you use to build it? It might look like a gigantic greenhouse for plants. What would you need to survive for a long time on Mars? What would you grow to eat? How would you cook? Who would govern, or run, the Martian community? What would you take if you were going to live in this ecosystem on Mars? Which of your friends and family would you want to join you? What kind of recreation, or fun stuff, would you be able to do to make life enjoyable?

If you desire, you can build your ecosystem on a piece of cardboard such as the one pictured below. Some things you could use are:

- Construction paper -- Use it to build houses and buildings.
- Salt dough (recipe is in the next project) -- Use it to make land and mountains.
- Sticks -- Use them to make trees. Place some crumpled green construction paper around them to make leaves
- Long, thin strips of cardboard -- Bend them over everything to make the habitat or ecosphere covering. You can then tape plastic cling wrap or wax paper to them to make large glass windows (see the photo below).

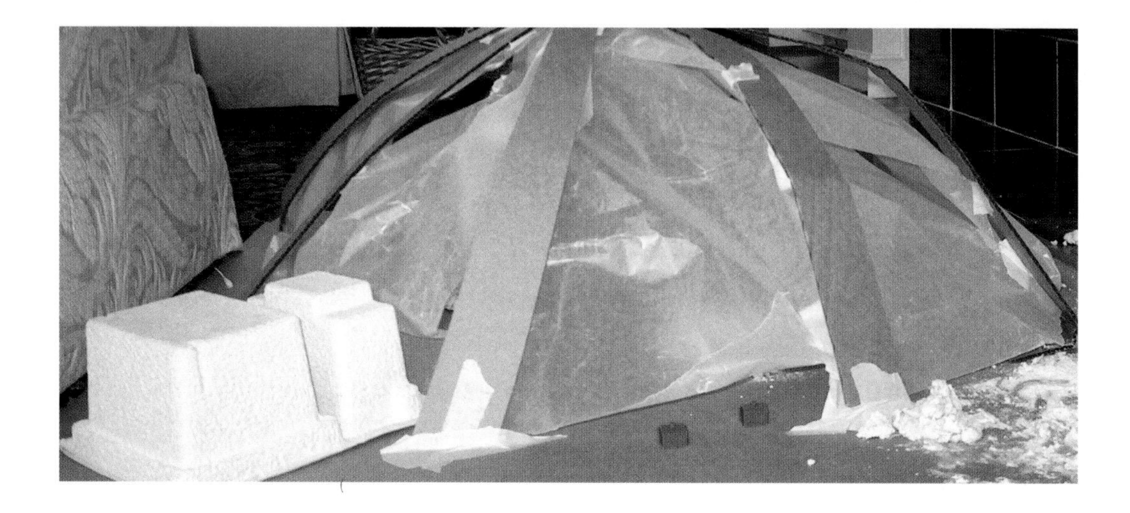

Younger Students: Tell the children to imagine they have taken a trip to Mars. Have them list everything they would need to take to Mars and who they would want to be there with them. Have them draw a picture of where they would live on Mars. If you have some extra boxes, allow them to build a spaceship for their travels to Mars. If you feel very adventurous, you can help them build an ecosystem for Mars like the older students.

Project
Build Olympus Mons

The biggest mountain peak in the entire Solar System is on Mars. It is a volcano called Olympus Mons. Here are two pictures of the volcano. One is straight above the volcano looking down, and the other is also above the volcano, but looking at it from one side.

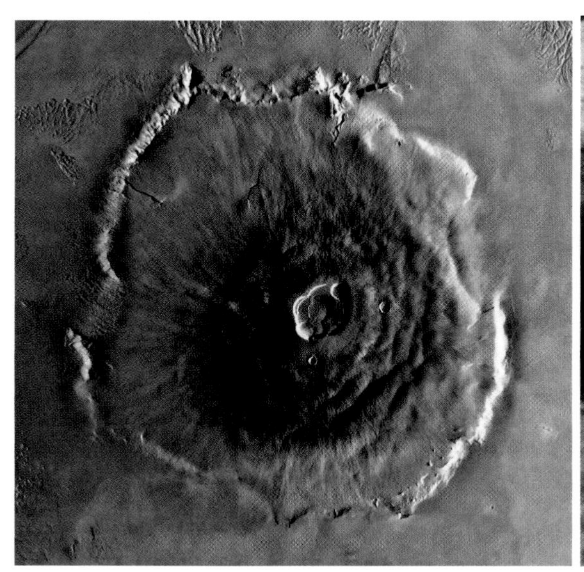

Olympus Mons from above

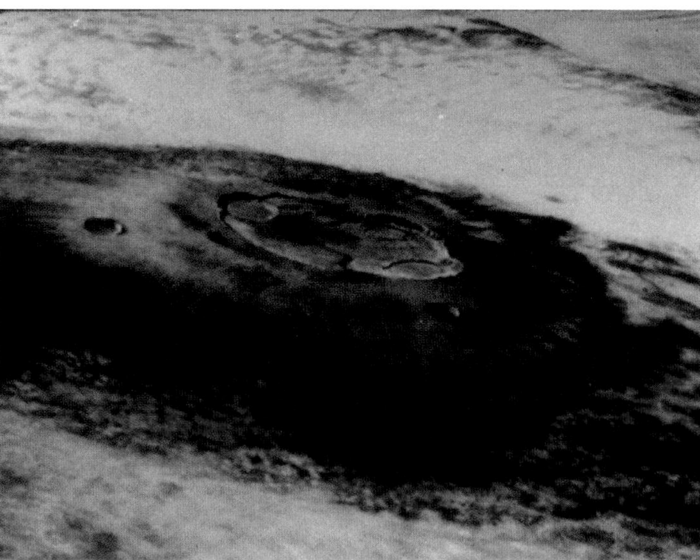

Olympus Mons from above and to one side

In this project, we are going to build a volcano that really erupts!

You will need:

- Salt dough recipe ingredients (see below)
- One small plastic bottle, such as a vitamin or aspirin bottle
- A large baking pan (not a cookie sheet) or aluminum foil
- Vinegar (White vinegar works best because it won't interfere with the color.)
- Rocks (optional)
- Baking soda
- Red food coloring
- Dish washing liquid

Salt Dough Recipe

In a large bowl, mix together:

1 ½ cup of white flour

¼ cup of salt

1 tsp. cooking oil

¼ cup of water
(Add tablespoons of water until it is a good dough consistency)

Instructions

1. Make the salt dough according to the recipe on the previous page.
2. Place your bottle in the large baking pan. The "lip" on the pan will keep the "lava" from spilling on the floor. If you do not have a large baking pan, you can lay out a large sheet of aluminum foil and then turn up the edges so that the sheet looks like a large baking pan.
3. If you have rocks, use them to surround your bottle, so that it looks like the bottle is on a rocky planet (see photo below).
4. Build a mountain around your bottle with the dough. Be sure to leave the top opening of the bottle uncovered, and be careful not to get dough inside the bottle. Take your time and make your mountain realistic. When you have a nice mountain, let it dry for the rest of the day. When it is dry, it will look like Olympus Mons, the biggest volcano in our solar system!
5. The next day, carefully pour warm water into your bottle until the bottle is about half full.
6. Add a few drops of red food coloring (or whatever color you think Mars's volcanic eruptions would be).
7. Put 2 drops of dish washing liquid into the bottle.
8. Add 2 teaspoons of baking soda to the bottle.
9. Pour some vinegar into the bottle and stand back!

You can experiment with different amounts of baking soda, food coloring, and dish washing liquid to see which eruption you like best.

Lesson 8
Space Rocks

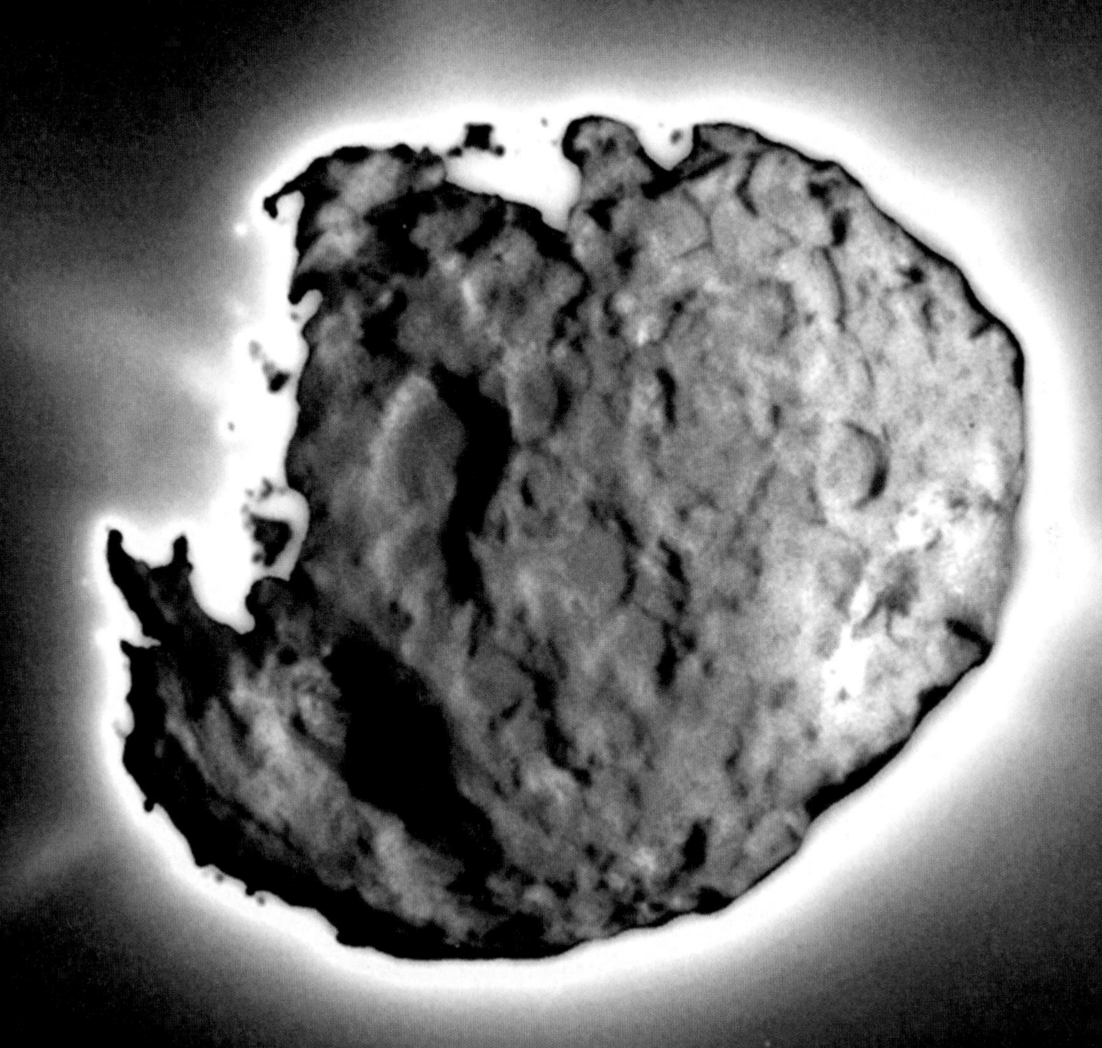

Millions of space rocks orbit our sun, but most of them are concentrated in an area between Mars and Jupiter. It makes sense, then, to have a lesson that discusses these space rocks before we study Jupiter, the next planet. The illustration on the front cover of this lesson is a NASA image of a comet that they call **Wild** (Vilt) **2**. It was taken by a spacecraft called "Stardust," the first NASA spacecraft designed specifically to study a comet.

Comets

This is a photo of comet Hale-Bopp in 1997.

Have you ever tried to build a snowman when there was only a little bit of snow on the ground? Well, if you have, you know how dirty that snowman looks because a lot of grass, dirt, and ground get rolled up into it. It becomes a big dirty snowball. Amazingly, that is exactly what astronomers call comets…dirty snowballs! Of course, comets are much bigger than your little snowman. In fact, many of them are so big that millions of snowmen could fit inside them.

The name "dirty snowball" gives you a clue about what a comet is made of, doesn't it? Comets are mostly big balls of ice. The ice is frozen water as well as dry ice, which you learned about in Lesson 7. This ice is mixed up with a bunch of rock, to make the center of the comet, which is called the **nucleus** (new' klee us). If you could make a comet, you would begin with a bunch of snow, ice, dirt, and rocks. Then, you would roll them all together into one big dirty snowball. After that, you would toss it into space to orbit the sun! It would be pretty neat to make a comet, but we know that only God can do that.

Comets have been seen for thousands of years. The ancient Chinese and other ancient civilizations kept written records of when comets appeared in the sky because they were frightened when comets appeared. Being unsure of what they were, many believed they meant something terrible was about to happen. Some of them appeared so bright that they could be seen during the day. Others had beautiful tails that would stretch halfway across the sky. You can imagine how strange that would have been to people who could see everything in the night sky so clearly because there were no city lights to drown out the stars.

This is a photo of comet Schwassmann-Wachmann 3 in 1995. Do you see how it looks like a big smudge in the sky?

The name **comet** comes from the Greek word **kometes**, which means "head of hair". A comet actually looks like a star with a head of hair, or a big smudge in the sky. If you ever got paint on your finger and tried to rub it off on a piece of paper, it would make a smudge. That is what a comet looks like to us. It's a bright, white smudge of light. It's really very large, though it looks kind of small.

The Coma

The reason a comet looks like a smudge in the sky is because it has a **coma** (koh' muh) around it. The coma is a big ball of steam surrounding the ice. If you have ever seen dry ice, you have seen steam coming off of ice. That is like a comet's coma.

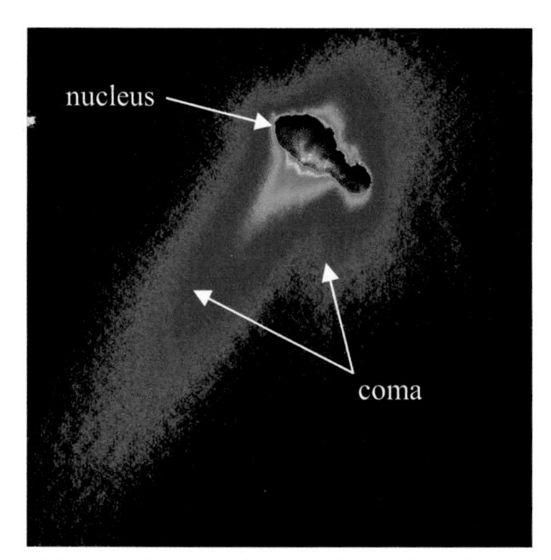

This is an image of comet Borrelly taken by the spacecraft Deep Space 1. False color has been used to emphasize the difference between the coma and nucleus.

When a comet first approaches the sun, it looks like a huge rock flying through space. Then as it gets close to the sun, it begins to heat up. The heat causes the ice to turn to steam, and that forms a steamy cloud around the comet, making the coma. The coma can be very large and bright, but it always looks a little blurry since it is really steam. As the comet flies through space, it leaves a tail of this steam behind it. Now remember, the comet is orbiting the sun. Do you remember what the solar wind is? It is a stream of particles shooting out from the sun. The solar wind pushes on the coma, and as a result, the coma always points away from the sun. Sometimes a comet has two tails; a yellow one and a blue one. The yellow tail is made out of dust particles, and the blue tail is the steam burning off the nucleus. The stuff in each tail reflects different light waves and produces different colors.

If you have ever watched Charlie Brown on TV, you might remember a character named Pigpen. A cloud of dust surrounds Pigpen, and when he walks, he leaves a trail of dust behind him. A comet is just like Pigpen! It always leaves a trail of dust particles behind it. These dust particles stay in the comet's orbit, so the comet has a very dusty, dirty orbit.

A Comet's Orbit

Comets can have long, elliptical orbits that are far outside of the planets. Some orbit close to the planets, passing by the earth quite often, and some have orbits that take them far beyond Pluto for thousands of years. A comet always follows the same orbit, so it is easy to predict when a comet will pass by earth again. Once a comet has been studied, scientists can predict when we will be able to see it again.

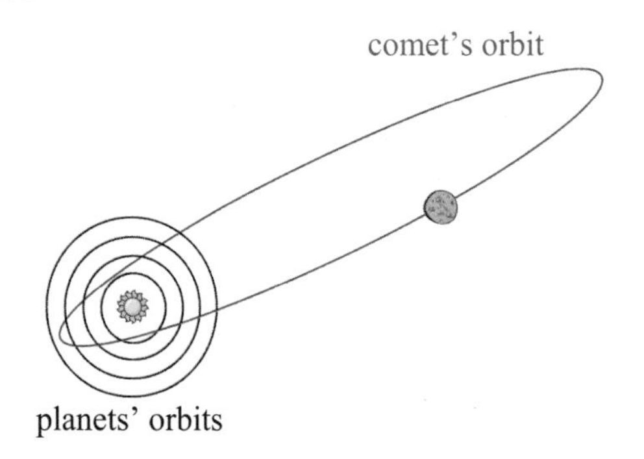

Remember that a comet is like Pigpen, leaving a trail of dirt behind it. When the earth passes through a comet's orbit, it runs into the dust and dirt that the comet leaves behind. An amazing and spectacular thing happens when these dust particles hit the earth's atmosphere. They catch on fire and create a little fireworks display for us! As they burn up, they look like beautiful shooting stars. It's a sight to see! These particles are actually very small, but, because they burn so brightly when they hit our atmosphere, we see them as if they are the same size as a star far out in space. Many people believe that a shooting star is an actual star up in space, but it's not. It's just a little piece of a dirty comet that left its mess behind when it passed by. When the earth moves into that dusty, dirty comet path, we see a bunch of shooting stars. When we have shooting star shows like that, we call them **meteor showers**.

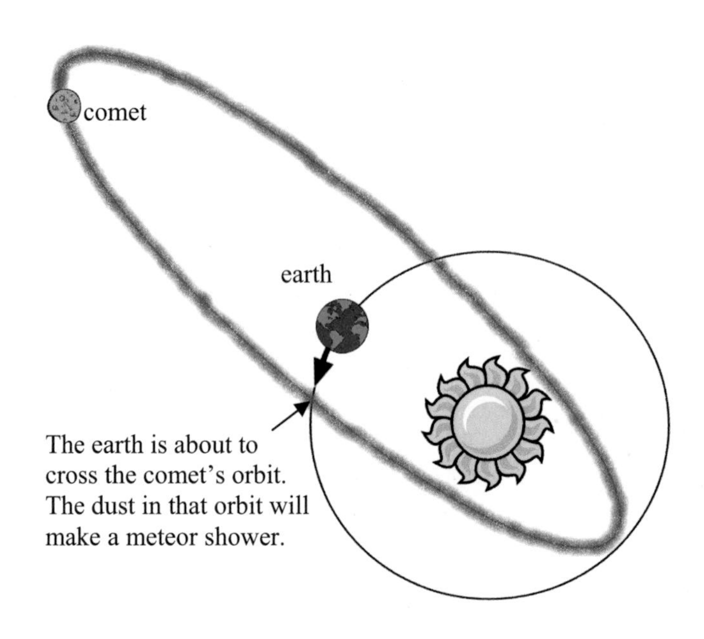

The earth is about to cross the comet's orbit. The dust in that orbit will make a meteor shower.

Because we know where the comets' orbits are, we know when we will pass through them. As a result, we can predict (which means to make a good guess) when we will have a meteor shower. There is a good one between August 9th and August 13th every single year. It is called the **Perseid** (per' see id) meteor shower. Mark your calendar now so you will remember! There are many more meteor showers each year. The activity at the end of this lesson involves watching the next meteor shower.

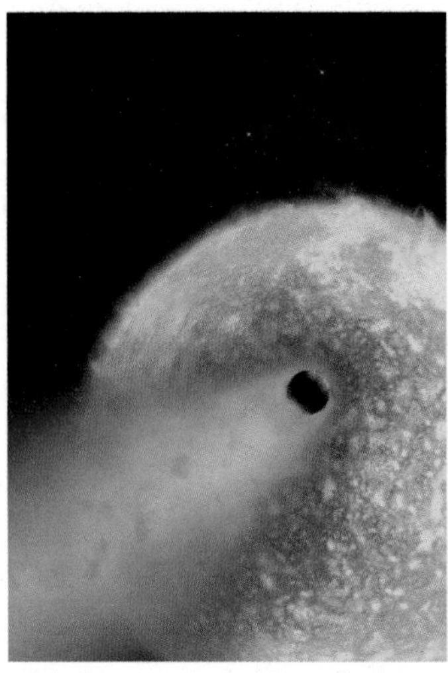

This is an artist's idea of what it looks like when a comet passes close to the sun.

If a comet's orbit around the sun takes less than 200 years, it is called a **short-period comet**. That means its orbit period is short. Now 200 years doesn't seem like a very short period to you and me, but for a comet, it is a short period. After all, compare that to **long-period comets** that can take thousands of years to make one orbit around the sun!

Most short-period comets get heated up by the sun so often that the ice around the comet gets burned off. It just goes by that hot sun too often to stay a big ball of ice for very long. To understand this, suppose you kept bringing a dirty snowball inside by the fireplace every few minutes. What would happen? It would eventually lose the snow and become just a pile of dirt. Since there is rock in all of that ice, when a comet's ice burns off, it looks like a big rock floating in space, called an **asteroid** (as' tuh royd).

Creation Confirmation

The fact that comets burn off their ice as they approach the sun is a problem for those who want to believe that the solar system is billions of years old. You see, if the solar system were billions of years old, the short-period comets that come by the sun so often would have no ice left on them by now! Since some comets come close to the sun every 70-80 years but still have ice on them, they can't be very old at all! In fact, comets must only be a few thousand years old since they still have ice on them. If they were millions or billions of years old, the ice would have all burnt off by now.

This is a picture of Halley's comet, which is a short-period comet, taking only 76 years to orbit the sun.

Of course, if you want to believe that the solar system is billions of years old, you can find some way to get around this problem. Those who believe that the solar system is very old think that there are sources in the solar system that continually produce new comets. They even have names for these sources: the Kuiper (ky' pur) Belt and the Oort (ort) cloud. They believe that the Kuiper Belt produces short-period comets and that the Oort cloud produces long-period comets. Although the Kuiper Belt does exist (you will study it later on in the course), it does not have nearly enough dirty snowballs in it to continually make new comets, so it doesn't seem like a good source for short-period comets. We have never even seen the Oort cloud, so we don't even know whether or not it exists. Even if it does exist, it probably doesn't have enough dirty snowballs, either. In the end, it seems more reasonable to believe that the solar system is only a few thousand years old.

Explain all that you remember about comets so far. Tell about how comets indicate that the solar system is only a few thousand years old.

Famous Comets

One famous short-period comet is known as comet Halley, or Halley's comet. This comet appears every 76 years. Chinese stargazers saw it 240 years before Christ was born! The Chinese kept great records of all they saw in the sky. They were very interested in the universe. In the 1600s, Edmund Halley studied ancient records and determined that the comet recorded by ancient Chinese astronomers returns every 76 years. Isn't it amazing that we can study ancient documents, written thousands of years ago, and learn more about astronomy? Sadly, Edmund Halley never got to see the comet's next arrival and learn that he was right after all. Seventy-six years is a long time. However, it did come by earth again, just when Mr. Halley predicted it would. Because of this, we named the comet after him. That was nice, wasn't it? Do you think you will ever have a comet named after you? What would we call the comet you discovered?

The biggest and brightest comet that has come by earth in the last century is comet Hale-Bopp. Its name is a little strange because two men discovered it, so it is named after both of them. This is a long-period comet, which means it will not return to earth for quite some time. It is so gigantic that its coma (the steam burning off of it) is as big as the sun. It didn't come close enough to the sun to cross our orbital path, so we do not get the pleasure of seeing a meteor shower display from its dust trail. Of course, it's probably a good thing that it did not pass too close to the earth. It is such a big comet, that a "close encounter" with it would be scary! After all, imagine what a huge comet would look like if it got too close to the earth!

Comet Hale-Bopp

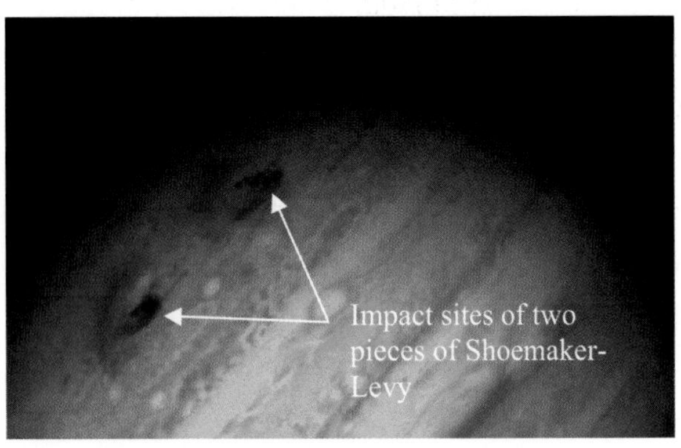

This Hubble Space Telescope image of Jupiter shows two spots where pieces of Shoemaker-Levy hit.

A comet that made recent history is the Shoemaker-Levy comet. This comet actually broke up into several comets. They were gigantic, and people watched them every night as they were orbiting the sun. One night, an amazing thing happened. As these big comet pieces were approaching Jupiter, they were pulled right into Jupiter's by its gravity. They crashed into Jupiter for six days, causing giant explosions on the gassy surface of Jupiter. It was an amazing sight to see.

If a comet comes by, be sure to take a good look at it, because it probably will not come back again in your lifetime. Even though you will not see the comet again, you may be able to see its trail every year when the earth passes through the dirty particles it left behind. Comets are beautiful lights traveling through the sky, and the trail of dust they leave behind gives us an amazing shower of lights. What a special gift of beauty God gave us! All of this reminds us of the glory of God and how He formed the world to show us how great He is. He gave us special gifts of lights in the sky that reflect His bright, shining glory.

This is a photograph of meteors during the Leonid meteor shower.

Meteorites

Remember all the dirt that comets leave behind? Dirt and rocks left by comets, as well as dirt and rocks from planets, are floating about in space. We call them **meteoroids** (mee' tee uh royds'). Each time a meteoroid hits the earth's atmosphere and burns up, it is called a **meteor** (mee' tee or). As you already learned, meteors burn up inside our atmosphere and look like little streaks of light, which we often call shooting stars. If a very large meteor hits our atmosphere, it will burn very brightly. These bright meteors are often called "fireballs."

Most meteors are very small. But occasionally, an object falls through the atmosphere and does not completely burn up because it is just too large. These larger objects hit the earth and are called **meteorites** (mee' tee uh rites'). No matter what it was while it was in space, if it hits the earth, it's called a meteorite. Usually the meteorites that hit fall into the ocean, since the earth is mostly covered by oceans. Once in a while, however, a meteorite will fall onto land. Some people are meteorite hunters, spending all their free time searching the earth for meteorites. The meteorites that have been found on earth have been many different sizes. There are some that are very small and some that are very large. The largest one found weighed 60 tons. That's as heavy as a whale, or thirty cars piled on top of each other.

This is a photograph of a meteorite that was once a part of Mars. It was probably launched from Mars as a result of an asteroid impact, and it ended up on earth.

There is a place in Antarctica where the ground is covered with meteorites. More than 10,000 (ten thousand) of them have been found lying on the snow. Antarctica is cold, and the ground there is always frozen with layers and layers of ice. Astronomers know that the rocks must be meteorites, because they were sitting above the thick snow covering the ground beneath. These rocks literally fell from the sky! Several of them were made of the same material found on Mars, so scientists think that they are Martian meteorites.

Asteroids

An asteroid is a rock orbiting the sun in our solar system. Wait a minute. Isn't a rock orbiting the sun called a meteoroid? Well, yes, it is. The difference between an asteroid and a meteoroid is size. Asteroids are large rocks (typically larger than a football field), while meteoroids are small rocks. Asteroids are not covered with ice the way comets are. Once a comet burns off all its ice, it looks just like an asteroid. This makes scientists think that at least some of the asteroids that are in our solar system are actually comets that have burned off all of their ice.

Asteroids are made of earth-like material like iron, rock, and carbon. When they enter our atmosphere, they begin to burn up, just as meteoroids and comets do. When an asteroid hits the earth, it is called a meteorite. However, it looks so much like the earth's rocks, that unless it falls in a place where there are not very many rocks, it is hard to tell it apart from other rocks. The ones that are found most often are the iron ones because they are shiny and stand out. There could be a meteorite in your backyard, but you probably can't tell the difference between it and an earth rock.

This is an artist's idea of what it looked like when the NASA spacecraft Deep Space 1 encountered an asteroid. This spacecraft tested new technologies that would be used in further deep space missions.

There are millions of asteroids out in space. Some are the size of a football field, and some are bigger than the state you live in. They come in all shapes as well. Many of them have been named, like Kleopatra, an asteroid that actually looks like a big dog bone! Asteroids are sometimes described as mountains in space or "planetoids," which means "little planets." Most of them can be found between Mars and Jupiter, in what is known as the **asteroid belt**. When a spacecraft goes to explore one of the planets past Mars, it must first pass through the asteroid belt. It can steer through the asteroids in the belt, however, because they are spaced far enough apart so that the spacecraft can wind its way in between them.

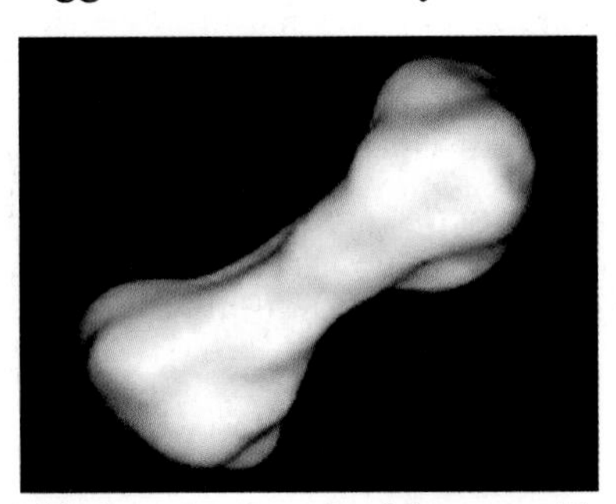

This is a photograph of the asteroid Kleopatra.

Can you tell me the differences between comets, meteoroids, meteorites, and asteroids?

Asteroid Belt

Let's learn a little more about the asteroid belt I just mentioned, because it is an interesting part of our solar system. The asteroid belt is a ring of asteroids that orbit the sun between Mars and Jupiter. We actually use the asteroid belt to separate the planets into two groups. Planets that are "inside" the asteroid belt (which means they are closer to the sun) are called the **inner planets**. Planets that our "outside" the asteroid belt (which means they are farther from the sun) are called the **outer planets**. This means that Mercury, Venus, earth, and Mars are the inner planets, while Jupiter, Saturn, Uranus, Neptune, and Pluto are the outer planets.

If you look at the drawing on the right, you will see that the asteroid belt looks a lot like the orbit of a planet. Well, there may be a reason for this. There is evidence that there was once a planet that orbited the sun between Mars and Jupiter. What happened to that planet? It may have exploded, and the asteroids in the asteroid belt may be the remains of that planet. Yikes!

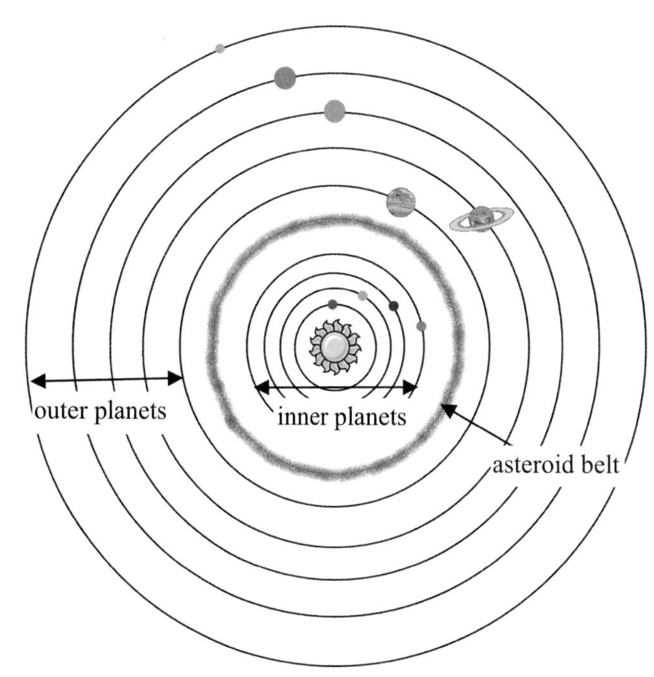

The asteroid belt separates the inner planets (Mercury, Venus, Mars, and earth) from the outer planets (Jupiter, Saturn, Uranus, Neptune, and Pluto).

The idea that there was once a planet between Mars and Jupiter is called the **Exploded Planet Hypothesis** (hi pahth' uh sis). A hypothesis is an idea that may or may not be true. Scientists make many, many hypotheses. That's what they do best! When something is a mystery to you, but you look at clues and then form an idea of what you believe, you have formed a hypothesis! For example, suppose you left a cookie on the table. If you came back to the table later and found your cookie was gone, you would have to form a hypothesis about what happened to it. Did someone eat it? Did Mom throw it away? Did Dad put it away? Perhaps someone opened the door, letting the wind blow in so strongly that it blew your cookie of the table, and then a giant ant came and took it back to its ant hole. Well, you can see that some hypotheses are good, and some are silly.

There are many reasons the Exploded Planet Hypothesis looks like a good hypothesis. If a terrestrial planet exploded, pieces that flew off into space would look like rocks and frozen water, or ice. That is exactly what asteroids and comets look like! Our planet is made of mostly water, so if it exploded, it would send millions of tons of water that would become ice out in space, perhaps forming huge comets. Scientists think that some of the comets and asteroids in our solar system are from this former planet that exploded. Did the water from this former planet form ice around the rocks when they were ejected into freezing cold space? We don't know for sure, but it gives support to the Exploded Planet Hypothesis.

More signs in favor of the Exploded Planet Hypothesis are the crater scars on many planets and moons. Remember that when a giant meteorite hits, it leaves a crater: a big dent in the surface. Some scientists believe that meteorites hit the planets and moons randomly over billions of years. But when one side doesn't have very many crater scars, it really bewilders them. Let me try to explain why.

Suppose you hung a big ball of clay from the ceiling and started it slowly spinning. Then, suppose you walked around and around the ball, throwing marbles at it for a year. If you did that, you would find dents all over the ball. It would be really strange if all the dents just happened to end up on one side, with none or very few on the other side. If you had ten balls spinning and you walked about throwing marbles at them for many years, the dents would be spaced out randomly over all the balls. The longer the balls hung while you threw marbles at them, the more likely it would be that craters would be evenly spaced all over the balls. On the other hand, suppose you threw marbles for only a few minutes. If you did that, there would not be enough marble dents to be evenly spaced on each ball. Most likely, they would be more concentrated on one side than the other.

Now suppose you had all ten balls slowly spinning around, and then one ball suddenly exploded into pieces that went violently and forcefully flying about the room. Most of the other balls would get hit, but they would mostly be hit on one side, the side facing the exploding ball. The balls closest to the exploded ball would probably have the deepest dents.

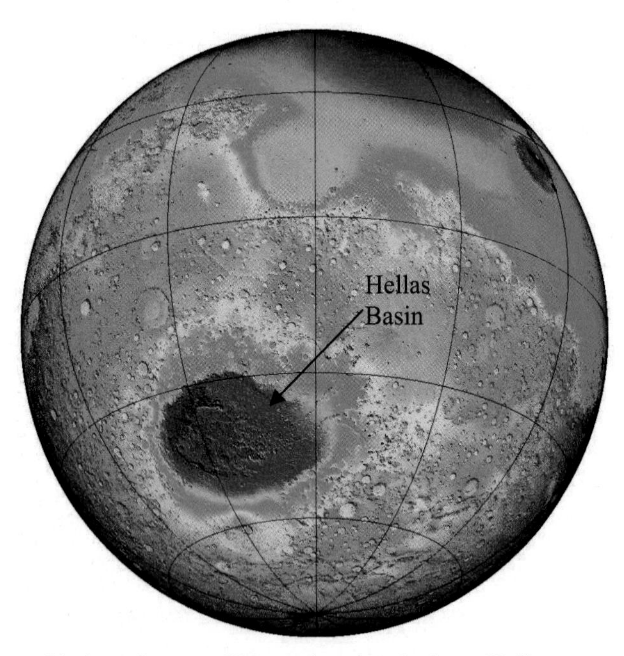

This is an image of Mars, showing the huge Hellas Basin. The colors are false; they are used to illustrate differences in the surface of the planet.

This is exactly what we see with our planets and moons! They have a great many more dents on one side than on any other side. This is true of earth and its moon, Mars, Mercury and many of the moons orbiting Jupiter, Saturn, Uranus and Neptune. In fact, the planet closest to the asteroid belt, Mars, has extremely deep craters. There is a crater on Mars (called Hellas Basin) that is bigger than the United States of America! If the asteroid belt was a planet that exploded, you would expect Mars to have had the most violent meteorite hits, resulting in the largest craters.

Do you recall that pieces of Mars have been found on the earth? We call them Martian meteorites. How did those get to the earth? If the planet next to Mars exploded, gigantic chunks of that planet would have hit Mars at such an outrageous speed that it would have caused pieces of Mars to launch into space.

Now, of course, if a planet exploded, pieces of it would certainly hit the earth as well. They would probably hit the earth at the same time that the Martian meteorites were hitting the earth, because they all formed as a result of the same explosion. Well, do you remember the thousands of meteorites I told you about in Antarctica? There are thousands of meteorites there all together, with only a few being from Mars. It's possible that most of those meteorites hit the earth as a result of the same event. We will never know for sure while we live on the earth how they got there, but it is the job of good scientists to find explanations for such things. The Exploded Planet Hypothesis might just be the correct explanation!

Some scientists say that the Exploded Planet Hypothesis couldn't be true because there are not enough asteroids in the asteroid belt to make up a whole planet. However, if we counted all the comets and other debris in our solar system that look like they came from a planet, this exploded planet would be much bigger than the earth!

If the Exploded Planet Hypothesis is correct, why did the planet explode? Some scientists think it was because a huge volcano erupted. Yellowstone National Park is built upon a gigantic, "super volcano." This volcano is so big that if it erupted, at least three states would have to be evacuated! See if you can find Yellowstone on a map of the United States (look in the northwest corner of Wyoming).

I doubt that the super volcano under Yellowstone will ever erupt in your lifetime, but if it did, it would mean big trouble. It would probably not cause the earth to explode, but it would cause a lot of damage. If a planet once existed where the asteroid belt is now, and if that planet had a super volcano that was even larger than the one under Yellowstone National Park, it is possible that such a large eruption could have caused the planet to explode.

Of course, if this planet once existed, there is another possible cause for its explosion. Some believe the planet between Mars and Jupiter exploded because a gigantic object, maybe a huge comet or asteroid, hit the planet. If a big asteroid or comet can cause huge craters such as the Hellas Basin on Mars, an even bigger asteroid or comet could cause a crater so deep that it would tear the entire planet apart!

This is an artist's idea of what it would look like if a large asteroid collided with a planet.

The asteroid belt is a very fascinating part of the solar system. It is a wonderful mystery that is fun to try and solve. The Bible tells us that right now, we see only a little bit of the truth, but when Jesus returns, we will see everything with perfect understanding. That will be a terrific time, and all of our questions will be answered. All the mysteries of the universe will be solved.

What Do You Remember?

What is another name for a comet? What does a comet leave behind it as it orbits the sun? What happens when a comet's dust particles enter our atmosphere? What do people call meteors? What is a meteor called when it hits the earth? Where have many meteorites been found? From which planet did some of the meteorites come? Where is the asteroid belt located? What is the Exploded Planet Hypothesis? Can you give some reasons that this might be a correct hypothesis?

Assignment

Get three sheets of paper. Title the first paper "Comets," the second "Meteors," and the third "Asteroids." Illustrate each page. Write down all that you remember about each type of space rock. Also, make a copy of the chart below and add it to your meteor page. Place everything in your notebook.

The following chart is a schedule of the meteor showers we have every year, but the dates are only approximate. To find out exactly when these meteor showers will take place, go to the course website I told you about in the introduction. There will be links there to help you find the exact dates.

Shower Name	Approximate Date to see	Name of Comet that left its dust (if known)	Number of shooting stars per hour
Quadrantids	January 3		40
Pi Puppids	April 5		40
Lyrids	April 22	Comet Thatcher	15
Eta Aquarids	May 5	Comet Halley	20
Delta Aquarids	July 30		20
Perseids	August 12	Comet Swift-Tuttle	50
Orionids	October 22	Comet Halley	25
Taurids	November 4	Comet Encke	15
Leonids	November 17	Comet Temple-Tuttle	15
Geminids	December 14	Asteroid 3200 Phaethon	50
Ursids	December 23	Comet Tuttle	20

Find the meteor shower that is closest to today's date. Then go to the course website and find the exact date or dates to view the meteor shower this year. Mark your calendar, and then go out and watch the show! Here are some tips on how to best view a meteor shower:

1. Do not use binoculars or a telescope. Your eyes are the best instruments to use when watching meteor showers.
2. Try to get away from the city lights. The city lights drown out the meteors. If you live in the city, see if you can go to a park or a country field far from the city lights.
3. Try to stay out later than you normally would. Typically, the best times to watch meteor showers are between midnight and dawn. If you can't stay up that late, just go out when it is very dark.
4. Bring a blanket or lawn chair that will recline. Lay on the ground or on the lawn chair and look up. Let your eyes adjust to the dark sky, and just keep watching for streaks of light in the sky. Those are the meteors.
5. Count how many you see. See if the number comes close to the number on the chart.

Notice that on August 12, the Perseid meteor shower (named for a constellation of stars in the sky at that time) gives us fifty shooting stars every hour. That is one you don't want to miss!

Project
Create a Scale Model Solar System

To fully understand the differences in sizes and distances of the planets in our solar system, we are going to gather some objects that represent the different planets and the sun. Many items can be found in your kitchen cabinets, especially where you keep the spices and the cookie decorating materials.

You will need:

- Sun - a ball, one that is about the size of a soccer ball
- Mercury - a tiny round sprinkle (dark colored), or a large grain of salt or sand
- Venus – a “red hot” (red cinnamon piece for cookie decorating) or a small, whole allspice
- Earth – another “red hot,” or small, whole allspice
- Mars – another round sprinkle, or large grain of sand.
- Asteroid belt – several sugar sprinkles that are smaller than the round sprinkle above
- Jupiter - a pecan or large marble
- Saturn - an acorn or large marble (a bit smaller than the Jupiter marble)
- Uranus- a pinto bean, or a lima bean
- Neptune – another pinto bean, or a lima bean (a bit smaller than the Uranus bean)
- Pluto- another round sprinkle (white)

Take these items and place them next to your sun to note the difference in size. These are the sizes of the objects in our solar system relative to one another. Wow! If you think you are amazed now, just wait! You will be even more amazed at how far apart they must be to make a true scale model of the solar system.

According to the sizes we have made the planets and the sun, we can calculate the distance using a yardstick. One yard (3 feet) will represent 3,600,000 miles, given the sizes of our planets. To be perfectly accurate, it would really be a little more than a yard, but one yard is easy for everyone to use. Even if you have a very large yard, you will only be able to place the inner planets at the scale distances, because it would take more than half a mile total to place all the planets in their orbit.

1. Mercury is about 36,000,000 miles from the sun. Because 3,600,000 is one yard and 36,000,000 is ten times 3,600,000, you will need to place Mercury 10 yards from the sun. You may want to tape Mercury to a piece of paper so you won’t lose it. You may want to do this for all of the planets.
2. Venus is about 67,000,000 miles from the sun. Beginning at the sun again, measure 18 yards to place Venus in its orbit.

3. Earth is about 93,000,000 miles from the sun, so you will place Earth 26 yards away from the sun.
4. Is your yard big enough to place Mars in its orbit? You will need to measure 40 yards from your sun, because Mars is 143,000,000 miles away from the sun.
5. The asteroid belt is about 280,000,000 million miles from the sun. That's 77 yards from the sun. I don't know if you can make it that far in your yard. Even the park might not be big enough for the asteroid belt to get a place in your scale model of the solar system.

I wish you could see the distances you would need to measure to get the outer planets into your solar system. It would help you to understand just how far away they are. Jupiter is 484,000,000 miles from the sun. That's 134 yards away from your sun - more than a whole football field! You can imagine how cold it is way out there. Saturn is 888,000,000 miles from the sun. That's a whopping 246 yards from your sun! Uranus is 1,783,000,000 miles (that's one billion, seven hundred eighty three million miles) away from the sun. Where could you go 495 yards to place Uranus in its orbit? Neptune is 2,796,000,000 miles from the sun (that's almost three billion miles)! You might get tired if you had to walk the 776 yards Neptune is away from the sun! Pluto is 3,673,000,000 miles from the sun. You would need to walk 1020 yards, more than ½ of a mile, to reach Pluto in your model. Can you imagine how far away that must be? Even as tiny as the planets are, that is how far they are from each other! It is simply amazing. Our God is a very big God, isn't He?

If you want to get a better idea of how the planets are spaced, we can use another scale for the distance. The scale will not be correct for the planet sizes that we are using, but at least this way, we will be able to put all of the planets into our model. In this scale, let's allow 6 inches to represent 36,000,000 miles. If we do that, here are the distances that the planets need to be placed from the sun:

1. Mercury: 6 inches
2. Venus: 11 inches
3. Earth: 1 foot, 3 ½ inches
4. Mars: 2 feet
5. Asteroid belt: 3 feet, 10 ½ inches
6. Jupiter: 6 feet, 8 ½ inches
7. Saturn: 12 feet, 4 inches
8. Uranus: 24 feet, 9 inches
9. Neptune: 38 feet, 10 inches
10. Pluto: 51 feet

Now remember, when you put your planets at these distances from the sun, the distances are not correct given the size of the sun and the planets. Nevertheless, it at least gives you an idea of the actual spacing of the planets in the solar system.

Lesson 9
Jupiter

Protective Mother

Jupiter is the first of the outer planets, and it is the giant of our solar system. It's the biggest planet that orbits the sun. It is so big that all of the other planets could fit inside of it. That's gigantic! If the earth were the size of a marble, Jupiter would be the size of a basketball.

Jupiter is like a protective mother to Earth. Just like God gave mothers to children for their protection, God also placed Jupiter in our solar system for our protection. God gave mothers love to keep children safe, and He gave Jupiter a large mass to keep the earth safe! You see, Jupiter is so massive that it has a very strong gravitational pull. Do you remember we talked about the rocks flying around in space? Meteoroids hit the earth and burn up, but comets and asteroids are much, much larger than meteoroids. We do not want any comets or asteroids flying into the earth! Thankfully, because Jupiter's gravity pulls so strongly, most of the comets and asteroids that do hit a planet end up hitting Jupiter instead of the earth. Do you remember the Shoemaker-Levy comet you learned about in Lesson 8? Its pieces hit Jupiter when it broke apart. They did not hit earth. Just think what might have happened had Jupiter not been there! We can thank God for making Jupiter so huge and far enough away that the comets do not fly into the earth. Comets and asteroids are just pulled right into Jupiter by its strong gravity. Life on earth would be very difficult if it were not for Jupiter's large gravitational pull!

This is a photograph of Jupiter. In a little while, you will learn about the Great Red Spot pointed out in the picture.

Going to Jupiter

Jupiter is the planet right after Mars, but if you wanted to go to Jupiter, you would have to pass through the asteroid belt in order to get there. That might be dangerous. When astronomers send spacecraft to Jupiter, it is a tricky thing to get the spacecraft past all those flying boulders.

If you got out of your spaceship and went down to stand on the planet Jupiter, you would sink right down to the very core of Jupiter. There isn't any ground to stand on! That's because Jupiter is not a terrestrial planet. It is not earth-like. It's what we call a **gas giant**, because most of it is made of gas, not rocks. It has a small, rocky core, but all around this small core are swirling gases. This means that Jupiter is mostly like a big ball of atmosphere. The atmosphere is definitely not pleasant, however. It's a very dangerous and hostile atmosphere. We definitely could not breathe if we went to Jupiter. Do you remember what chemical we need in the atmosphere to breathe?

Little Sun

Jupiter is too far away from the sun to get much heat from it. As a result, it is a cold planet. However, it's not as cold as you might think, because Jupiter has its own heat source. Remember, Jupiter has a small, rocky core. Well, it turns out that this core is very hot, and it is cooling off by leaking heat into the gases that make up most of Jupiter. This actually makes Jupiter warm enough to leak heat into space. Instead of just absorbing the sun's heat, then, Jupiter also puts out its own heat! Although the heat put out by Jupiter is tiny compared to the heat put out by the sun, the fact that Jupiter puts out heat makes it a little like the sun.

Jupiter has something else that makes it like the sun. Jupiter is made mostly of two gases called **hydrogen** (hi' druh jen) and **helium** (he' lee uhm). You might recognize helium; it is the gas that makes balloons float in the air. If you let go of a balloon filled with air, it just falls to the ground. If you let go of a balloon full of helium, it floats up into the sky. Well, it turns out that the sun is made mostly of hydrogen and helium as well. So, Jupiter is a little bit like the sun. It puts out heat like the sun, and it is made mostly of the same gases as the sun. It even has satellites like the sun. The sun's satellites are the planets. Jupiter's satellites are its moons, which you will learn about in a little while.

Stormy Skies

If you look at the picture on the first page of this lesson and the picture on page 104, you will see a big spot on Jupiter. That's the **Great Red Spot**, and it is a giant storm (like a hurricane) that travels around the planet. This storm is many times stronger than the worst hurricane on earth, and it has been raging on Jupiter for more than 300 years! How do we know this? Astronomers have seen it in their telescopes since at least 1665. The Great Red Spot is twice the size of the earth! That is one huge, long-lasting storm.

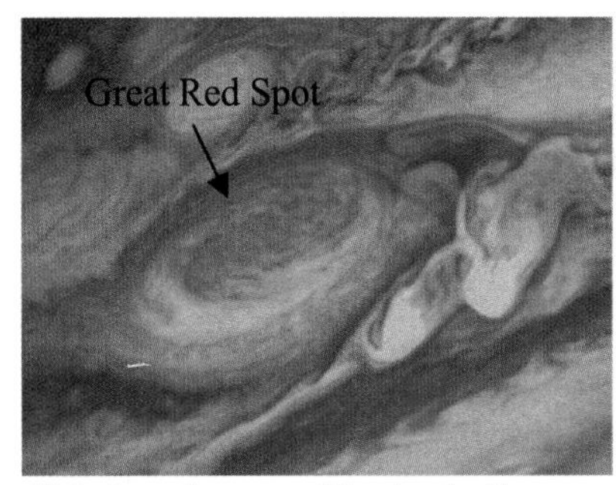

This is a closeup of Jupiter's Great Red Spot.

In pictures of Jupiter, we see beautiful stripes across it. However, those stripes aren't all that beautiful when you get up close. Those stripes are actually clouds that hold violent storms. Jupiter has lightning that is much more powerful than the lightning on earth. With all of those storms and lightning going on, Jupiter seems like a very frightening place doesn't it? It's hard to imagine how we could ever send people to Jupiter. God did not make Jupiter a place that would be safe to explore.

Tell someone everything you have you learned about Jupiter so far so that you won't forget it.

Jupiter's Rings

We have always known that Saturn has rings, but astronomers have recently discovered that Jupiter has rings around it, too! These rings are not visible through telescopes, because they are very thin, and the light that Jupiter reflects from the sun makes them even harder to see. In fact, they were not seen until a spacecraft got close enough to take a picture of them. We usually don't illustrate Jupiter with its rings since you cannot see them with a telescope.

This is a picture of Jupiter's rings as taken by the spacecraft Galileo. Jupiter cannot be included in the picture, because its light from Jupiter would drown out the light being reflected by the rings.

Rotation and Revolution

Jupiter rotates very quickly. It takes only ten hours for Jupiter to rotate one time. That means that a day on Jupiter is less than half of an earth day. Nighttime on Jupiter is only 5 hours, and daytime is 5 hours as well. Those are short days and nights! Do you remember from Lesson 5 that the speed of a planet's rotation affects the weather? Earth's rotation is slow enough that the weather is not very violent. Jupiter's fast rotation is part of the reason there are violent winds and storms there.

Jupiter is further away from the sun, so a revolution around the sun takes much longer. It takes Jupiter twelve earth years to orbit the sun. We call that one Jovian (joh' vee un) year. The word "Jovian" refers to Jupiter, because Jupiter was named after the Roman god of rain, thunder, and lightning. Another name for this false god is "Jove," so "Jovian" refers to Jupiter. Since one Jovian year is twelve earth years, when you turn twelve on the earth, you would be only one Jovian year old. You would be a big one-year-old wouldn't you? How old would you be on Jupiter if you were 48 earth years old?

Many Moons

Do you remember that a moon is a satellite? Well if Jupiter is like a mom, she has a lot of children! Jupiter has more than 60 satellites. In some ways, Jupiter is like its own miniature solar system! It seems that most of Jupiter's satellites are adopted. Jupiter's gravity is so strong that the planet can pull many objects into its orbit. Most objects pulled into Jupiter crash right into the planet, but it appears that some of Jupiter's moons got pulled in and didn't crash. Instead, they became satellites.

The four biggest satellites, or moons, orbiting Jupiter are named **Io** (ee' oh,) **Europa** (yuh roh' puh,) **Ganymede** (gan' uh meed,) and **Callisto** (kuh lis' toh). Galileo discovered these moons many years ago. Do you remember who Galileo was? He was a famous astronomer who taught us a great deal about space. He was the first to use a telescope to seriously study the heavens. In honor of Galileo's discovery, these four moons are called **Galilean moons**.

This is a size comparison of the Galilean moons and Jupiter.

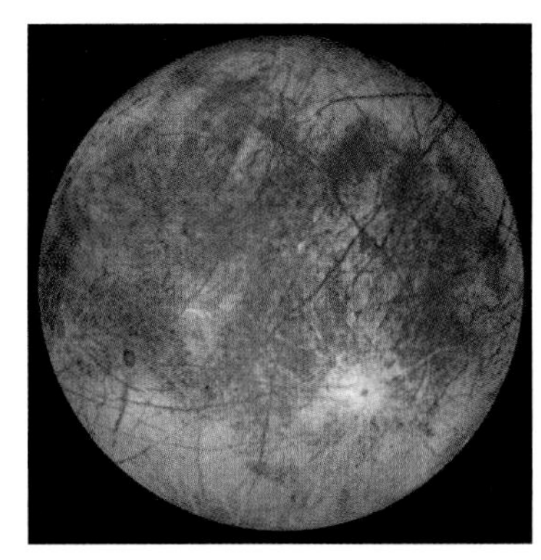
Europa – an icy moon

Europa is the smallest of the Galilean moons. It does not have craters on it like the other moons. Astronomers are amazed at how smooth Europa is. In fact, it is the smoothest object in the solar system. Most astronomers think that Europa is covered with a large, frozen ocean. Do you see all of the lines in the photograph of Europa? Those are cracks in the ice. Evidence from the Hubble Space Telescope tells us that there is oxygen on Europa. However, don't think that you can go there and breathe. The amount of oxygen on Europa is so small that you could not survive if you tried to breathe it.

Io is a strange name isn't it? It is the second smallest of the four Galilean moons. Io has hot, active volcanoes, but they aren't filled with lava! They are filled with a sulfur chemical that smells like rotten eggs, so I'm sure Io smells really bad! Every once in a while, NASA astronomers see the sulfur chemical spewing into the atmosphere of Io, and that's what gives Io its interesting color. In fact, Io is the most colorful moon in the solar system.

Io – a colorful moon

Callisto – a ball of ice and rock

Jupiter's second largest moon, Callisto, is about the same size as the planet Mercury. Astronomers recently learned that Callisto is a strange moon, indeed. It doesn't have any core at all, the way most moons and planets do. It is a huge ball of ice with rocks and boulders scattered about in the ice. It also doesn't have a hot inner core like the earth's moon. It's very much like the nucleus of a comet. It is also very similar to Pluto. Interestingly enough, Callisto has one of the largest impact craters in the solar system. It is called "Valhalla," and it is about 373 miles across. That must have been one big asteroid or comet!

Ganymede is Jupiter's biggest moon. In fact, it is the biggest moon in our whole solar system. It's bigger than the whole planet Mercury. It would be considered a planet if it were circling around the sun instead of Jupiter. It is similar to Callisto in the sense that it is made of ice and rock, and it is covered with craters. Unlike Callisto, however, it is not just a simple mixture of ice and rock resembling a comet's nucleus. Instead, the moon has a definite structure. There is a hot core made of iron and sulfur, a rocky covering around the core, and then an icy covering over all of that. Like Europa, there is some oxygen on Ganymede, but it is definitely not enough for you to breathe if you wanted to visit there.

Ganymede – a giant moon

It seems that many of Jupiter's moons are really gigantic balls of ice. Some believe these frozen ice balls could have been comets pulled into the gravitational pull of Jupiter, becoming satellites.

This is a photograph comparing the size of Amalthea (upper right) to Io.

Amalthea (um al thee' uh) is a very odd Jovian moon. It is a jumble of gigantic rocks pulled together by gravity. Suppose you grabbed a bunch of rocks from your yard and glued them together to make a rock structure. Your rocks wouldn't fit perfectly together because they were not made to go together. That's like Amalthea. It has big spaces between the boulders, just as your rock structure would have. Your rock structure probably wouldn't be round. Almathea isn't round either. It is also small compared to the Galilean moons. Notice how small it is compared to Io, the second smallest Galilean moon. Amalthea might be pieces of the exploded planet that might have formed the asteroid belt.

What do you know about Jupiter's moons? Explain what you have learned.

Spacecraft Galileo

Almost everything we know about Jupiter comes from the unmanned spacecraft called Galileo. In 1989, a rocket blasted Galileo out of the earth's atmosphere and let it go. Galileo went on an interesting journey to Jupiter. It took six years to get to there. It would not normally take that long to get to Jupiter, but this spaceship took the long way around.

When Galileo left earth, it went the opposite way, toward Venus. It did this because it needed to use Venus' gravity to swing it around and propel it into outer space. This saved fuel, because Venus's gravity flung the spacecraft towards Jupiter. When it flew past the earth toward Jupiter, it got a big swing from the earth's gravity, too. You see, when a spacecraft gets close to a planet, the planet's gravity grabs the spaceship. With all the speed coming from the spaceship, the pull of the gravity, and the circular motion of the orbit around the planet, the spaceship gets a big push into outer space. NASA engineers often use the gravity of planets to propel spacecraft into outer space. They call it a "gravity assist."

While Galileo was flying past the earth, it took many beautiful pictures. Then it passed the orbit of Mars. Mars was on the other side of the sun at the time, so Galileo didn't get any pictures of Mars. After it passed Mars's orbit, guess what it ran into? It ran into the asteroid belt. It didn't hit any asteroids and no asteroids hit it. Phew! That was a good thing. Galileo did take some pictures of asteroids it passed. Since it was traveling on a curved path, Galileo actually came back to earth and got one more big push into outer space. After that, it was finally on its way to Jupiter. Of course, it had to pass through the asteroid belt again, but it made it through without any collisions, so it got to Jupiter okay. It took many beautiful photos that were sent back to earth by a computer. Most of the pictures in this lesson come from Galileo.

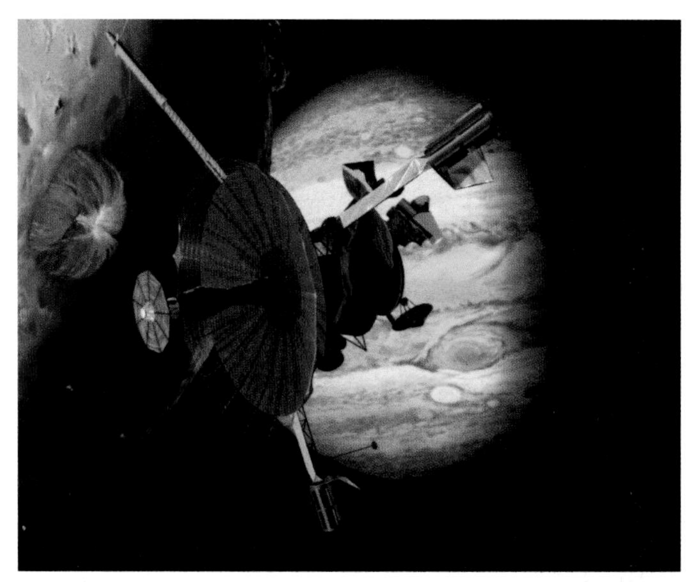

This is an artist's idea of what it looked like when Galileo reached Jupiter and started studying Io.

When Galileo got to Jupiter, it sent a little robot into the middle of Jupiter to get information. It's really amazing that the probe was able to survive in Jupiter's harsh and dangerous environment; but it did, and it sent back information to Galileo about Jupiter.

Galileo had one big problem when it got out into space. Its big antenna would not open! NASA needed a really long antenna so that when Galileo was all the way across the solar system at Jupiter, they could still receive information from it. That's too far away for a little antenna to send a lot of pictures and information. The astronomers down on the earth tried everything to get the antenna to come up. They tried pointing Galileo into the sun; they tried turning it away from the sun; they even tried turning Galileo on and off. They tried everything, but nothing would work. Finally, they had to learn to make due with the little bitty antenna, so they did not get as much information as they could have gotten had the large antenna worked. When Galileo finished its mission on Jupiter, scientists programmed it to crash into Jupiter, disappearing beneath the foggy surface, into the unknown.

Finding Jupiter in the Night Sky

Jupiter is the second brightest planet in our solar system. Jupiter is so large that you can see some of its details with just a pair of binoculars or a small telescope. Sometimes, you can even see Jupiter's moons with just binoculars or a small telescope! The way to see the moons with binoculars is to steady the binoculars against something solid. You could also set them on something, like a car or a mailbox. This will keep them focused on Jupiter. After your eyes have had a minute to adjust to the light, you will see one or more small specks of light around Jupiter. Those are Jupiter's moons!

If you need help finding Jupiter in the sky, visit the course website I told you about in the introduction to the course. It will have links to websites that will help you find Jupiter.

What Do You Remember?

Tell someone all that you can remember about Jupiter. How does Jupiter protect our planet? Why is Jupiter a little like the sun? What is the Great Red Spot on Jupiter? What do you remember about that spot? Why does Jupiter have stripes? Name Jupiter's largest moons. Why are they called Galilean moons? Can you describe Amalthea? What do you remember about the spacecraft Galileo?

Assignment

Make a Newspaper

Older Students: For your notebook, you are going to do something fun. You are going to make the front page of a newspaper. The newspaper will be all about Jupiter. Get a newspaper and study the front page. An example of one is shown on the next page. Try to make your layout like a real newspaper. The name of the newspaper should go across the top. Name your newspaper something catchy. Below the title, there should be a lot of different headlines. The headlines are the bigger letters that tell you what the story below is going to be about. Newspaper reporters try to make the headlines exciting so that you will want to read the story they wrote. The story they write below the headline is called an article.

After you have written the name of your newspaper across the top, put today's date in smaller writing below the title. Make the headlines about Jupiter: the storms, its moons, or the spaceship Galileo. Write a short article below each headline. You might want to begin with a sketch (a pencil drawing) of what you want your newspaper to look like. Put your finished paper in your notebook

Younger Students: Tell your child that he is going to write a newspaper about Jupiter. Explain to him that he will be pretending to tell other people about Jupiter. Show him a newspaper and its features and layout. There is an example of one below. Draw a layout on a piece of paper as an example and then have him do one on his own. Have him decide what he wants his newspaper articles to say about Jupiter. Help him come up with a catchy name and interesting headlines that jog his memory about what you read to him. Have him remember what he learned about Jupiter and write those things in each of his articles.

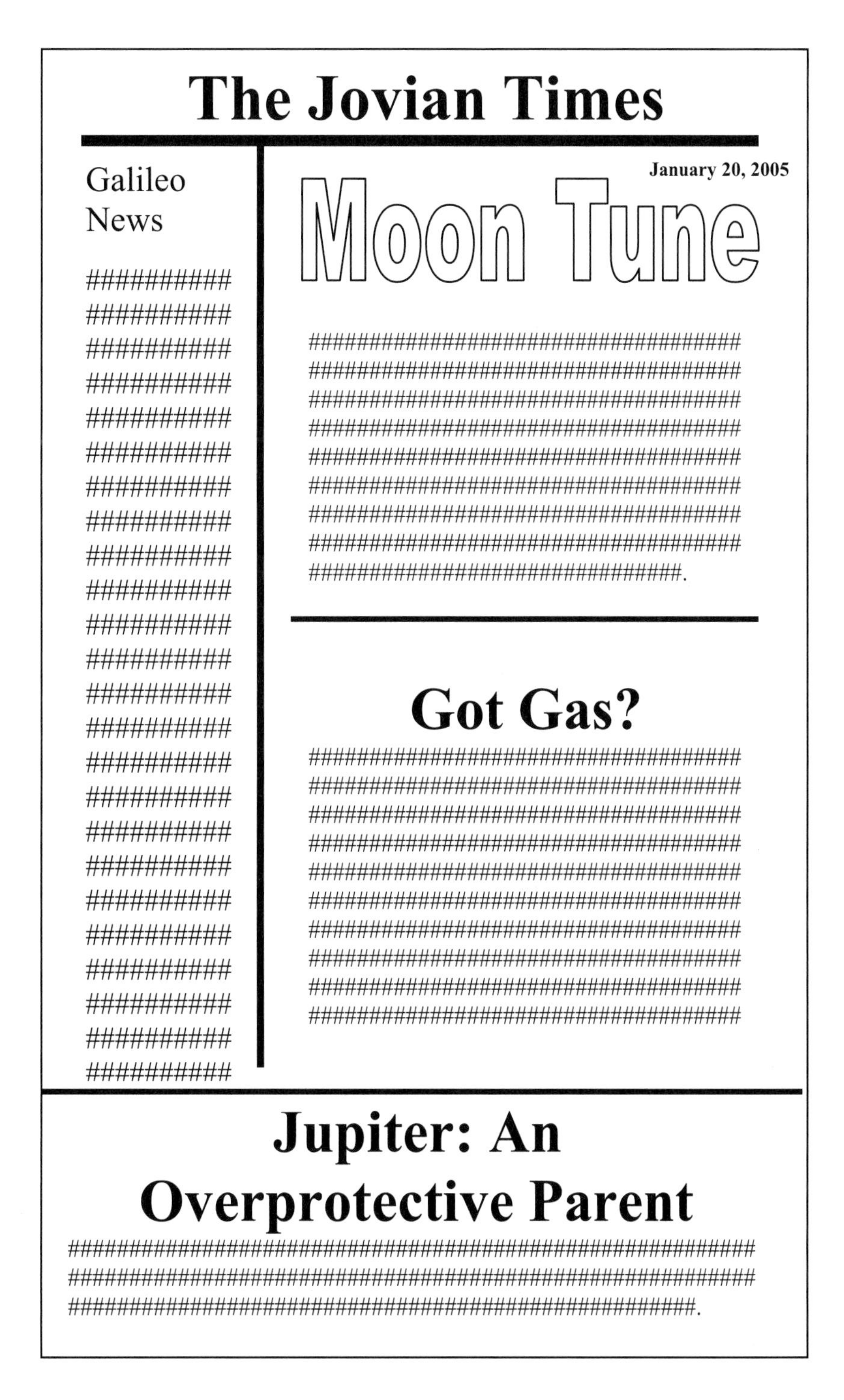

The Jovian Times

January 20, 2005

Galileo News

##########
##########
##########
##########
##########
##########
##########
##########
##########
##########
##########
##########
##########
##########
##########
##########
##########
##########
##########
##########
##########
##########
##########
##########

Moon Tune

##
##
##
##
##
##
##
##
###################################.

Got Gas?

##
##
##
##
##
##
##
##
##
##

Jupiter: An Overprotective Parent

##
##
##.

Project
Make a Hurricane Tube

The Great Red Spot on Jupiter is like a hurricane on earth. We are going to make a hurricane in a bottle as an example of what a hurricane on Jupiter might look like.

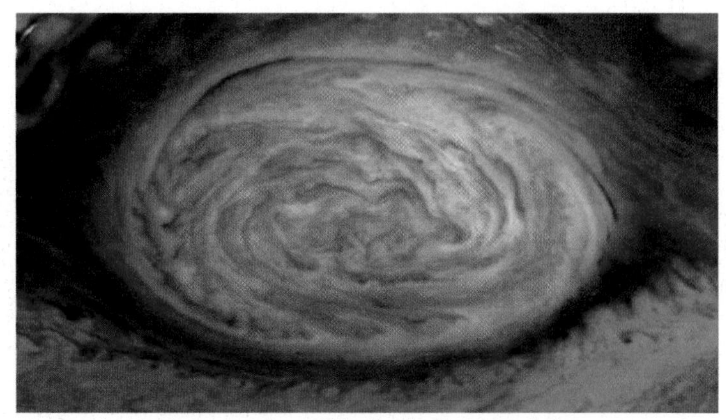

A close-up view of Jupiter's Great Red Spot

You will need:

- ♦ Two plastic soda pop or water bottles. (They can be large or small. The larger ones give you a bigger hurricane.)
- ♦ A 1-inch washer
- ♦ Water resistant tape (Electrical or duct tape works best.)

Instructions

1. Fill one bottle two-thirds full of water.
2. Place the washer on top of the mouth opening (see the picture on the top right).
3. Place the other bottle upside-down on top of the washer.
4. Tape the two bottles together very thoroughly (see the picture on the bottom right). If your experiment leaks, you will need to add more tape and wrap it tighter.
5. Turn the two-bottle setup over so that the bottle with water in it is now the upside-down bottle that is on top.
6. Move the bottles in a circular motion, swirling the water around in the same direction for a moment.
7. Watch your system form a hurricane-shaped funnel down into the bottle below. You can do this over and over again. Some people like to put glitter or tiny toys in the hurricane tube so they can watch them spin down into the tube below.

Lesson 10
Saturn

Saturn

Saturn is the most famous planet in our solar system. It even has a day of the week named after it…Saturday! Everyone remembers what the sixth planet from the sun looks like, because it has bright rings that circle around it.

Saturn is easy to see in the sky, so it has been studied throughout history. However, the rings around Saturn were not discovered until 1610, when Galileo looked at Saturn through his telescope. Because of the weakness of Galileo's telescope, the rings looked like two little handles sticking out on either side of Saturn. Thinking they were cup handles or odd-looking moons, he was very confused about them. Forty-five years later, another scientist named Christian Huygens (Hoy' ginz) used a more powerful telescope to study Saturn, and he realized that they were rings!

Saturn is tilted as it rotates. In fact, its tilt is similar to earth's tilt. Do you remember what the earth has because of its tilt? It has seasons. Since Saturn is tilted like the earth, it also has seasons. Saturn's tilt causes something else as well. It causes the view of Saturn from the earth to change. During part of its trip around the sun, we see mostly the edges of Saturn's rings. As it continues in its orbit, however, our view of Saturn changes, and its rings are much more visible.

This picture shows the view of Saturn from earth during different times in Saturn's orbit around the sun.

Saturn is a little smaller than Jupiter, making it the second biggest planet in our solar system. Saturn and Jupiter together are the largest gas giants. The next two planets further out, Uranus and Neptune, are also gas giants, but Jupiter and Saturn are much larger. To see the difference in size between the earth and Saturn, get a marble and a soccer ball. The soccer ball would be Saturn, and the marble would be the earth.

Twins

God made Saturn a lot like Jupiter. They are almost like twins, and astronomers often call them twin planets. Saturn is a gaseous planet. Do you remember the difference between a terrestrial planet like the earth and a gaseous planet like Jupiter or Saturn? Just like Jupiter, Saturn is mostly made of hydrogen and helium gas.

Being further away from the sun, Saturn is much colder than Jupiter. Saturn is usually about 300 degrees below zero! The very center of Saturn, however, is hot, hot, hot, with a temperature greater than 20,000 degrees! Like Jupiter, Saturn puts out its own heat, rather than getting all of its energy from the sun.

Saturn is so beautiful, but her beauty is deceiving, for Saturn is a horridly cold and stormy planet. Although Saturn's storms are not quite as violent and long-lasting as Jupiter's storms, you still wouldn't want to be caught in one! Do you see the stripes that go around the planet? Those are bands of clouds, and the storms occur within those clouds. The stripes add to Saturn's beauty, but the beauty of those stripes to us here on earth would be trouble for us if we visited Saturn! The winds within those stripes blow fiercely at about one thousand miles per hour. That's five to ten times stronger than the strongest hurricane on earth! We couldn't survive if the earth had winds that strong blowing on it. That's why God created the earth differently than the other planets. He didn't create Saturn as a nice place for people to live or vacation.

This is a photo of Saturn. The small white spots are two of Saturn's moons.

Saturn is the only planet in our whole solar system that is less dense than water. What does that mean? Well, if something is more dense than water, it will sink in water. Jupiter would sink if it were placed in water, because it is more dense than water. If something is less dense than water, it floats in water. So, if there were a body of water large enough in which to place Saturn, it would float! Saturn is the only planet that would float in water. Are your bathtub toys more or less dense than water?

Ring System

There are thousands of rings around Saturn. Yes, there are thousands! It looks like one or two rings when you look at Saturn through a telescope on earth, but spacecraft have gotten close enough to see that there are many more than that. The rings are made of dust, ice, and rocks that orbit around the planet. Some of the rocks are the size of a little pebble, and some are the size of a whole house! If you could stand on Saturn, you would see big, giant, house-size chunks of rock and ice swirling above you all of the time. Astronomers aren't completely sure what formed these rings. Some think that giant asteroids or comets crashed into one or more of Saturn's moons, smashing them to pieces and making the rings. Others think that the rings were once asteroids or comets themselves, but they got so close to Saturn that they were broken

The colors are false in this photo of Saturn's rings. They are used to show you that there are lots of rings.

into tiny pieces, making the rings. Saturn's rings are one more neat mystery that God has given us. Perhaps you will one day figure out exactly what made them!

This is an artist's idea of what Saturn's rings look like. The shepherd moons are not pictured.
Digital artwork by Dr. David Heatley

Saturn's rings also contain whole moons. What are those moons doing in the rings? Well, do you know what a shepherd does? A shepherd takes care of sheep, keeping them all together and making sure they go where they should. Saturn has **shepherd moons**, because they herd in the rings as if they were sheep, keeping them from spreading out too far. The two best-known shepherd moons are called Pandora (pan dor' uh) and Prometheus (pruh me' thee us).

Saturn's rings are so thin that sometimes you cannot even see them with a telescope. Remember, the view of Saturn from the earth changes, and if the tilt is just right, we can only see the edges of the rings. The edges are only about 50 miles thick, which is tiny for a gigantic planet like Saturn. That makes them hard to see in some telescopes. Although they are thin, they are very wide. Because they are wide, when Saturn's tilt is right, we can see the rings quite well with a telescope.

Saturn is even further from the sun than Jupiter. It takes almost 30 earth years for Saturn to orbit the sun just one time. If you were two Saturnian years old, then, you would be sixty years old on earth. You might even be a grandparent!

What do you remember about Saturn so far? Explain in your own words all that you have learned.

Fast Rotation

Have you ever spun a top? A top twirls around extremely fast. Saturn is spinning around so fast that it's daytime for just over 5 hours, and then it's nighttime for just over 5 hours. Guess what would happen if you went to bed on Saturn and slept for 10 hours, which is how long most children sleep? When it was about time to wake up, it would just be getting dark again! Around lunch, it would begin getting light again. When you were ready to go back to bed, it would get dark again. It would just keep getting dark, then light, then dark, then light over and over again every five hours. Saturn is spinning so fast that it makes the planet bulge at the middle and flatten at the top. This makes Saturn look like somebody is squashing it. Do you remember that Jupiter rotates quickly, too? Well, Saturn's rotation is just a little slower than Jupiter's. Do you remember that Jupiter's fast rotation is one reason it has such strong winds and terrible storms? Well, the same is true for Saturn. Since Jupiter spins a little faster than Saturn, Jupiter also looks like someone squashed it. The fast rotation of these two gas giants is just one more reason that they are considered twin planets.

Saturn's Moons

This is a picture of Saturn and its seven largest moons.

If you spent the night on Saturn, it wouldn't be very dark, because astronomers have discovered more than 30 moons that orbit around Saturn! That means Saturn has more moons than any other planet except Jupiter. You can see some of them in the picture at the beginning of this lesson. They are the tiny white circles to the left of Saturn.

The biggest moon is named **Titan** (tie' tun), and it is almost as big as Mercury. Titan looks orange from a spaceship. It gets that color from a thick atmosphere that is full of clouds. These clouds make it hard to see the surface of the moon from a spaceship. The seven largest moons of Saturn are shown in the picture above. The relative sizes of the moons are not correct, because the artist who put the picture together made it look like you are closest to the moon Dione (dy oh' nee). Many of Saturn's moons are small and look like big, irregularly-shaped rocks rather than nice round moons. Most likely, they are space rocks that got caught in Saturn's gravitational pull.

Cassini Mission

This is an artist's idea of what Cassini will look like orbiting Saturn.

As I am writing this book, there is a spaceship traveling to Saturn to explore the planet and its moons. When a spacecraft travels somewhere, it is on a "mission." This mission is called the **Cassini** (kuh see' nee) **Mission**. Cassini is a very big spacecraft. It is taller than a two-story building, and it weighs more than 6 tons. That's heavier than a large moving van. It took a lot of rocket power (and money) to blast it into space. Cassini will orbit Saturn for four years, taking pictures of Saturn and her moons. It will send a probe named **Huygens** to collect and test samples of dirt from Titan. It is supposed reach Saturn in July of 2004. If you go to the course website I told you about in the introduction, you will find a link that will let you follow the progress of the Cassini mission.

Finding Saturn in the Night Sky

Saturn is easy to see in the night sky. It isn't as bright as Jupiter, but you can tell it apart from the stars because the light is steady and not twinkling. If you have a telescope, you can even see the rings of Saturn. During certain years, Saturn is tilted on its side so that you can see the rings perfectly. If you go to the course website, you will find links to places that will help you to find Saturn.

What Do You Remember?

Be sure to tell someone all that you remember about Saturn. What is Saturn made of? Why would Saturn be an unpleasant place to visit? Which planet is considered Saturn's twin? What are Saturn's rings made of? What do shepherd moons do? How many years does it take Saturn to orbit the sun? Why does Saturn look as if it is being squeezed? What is the name of the space mission that is going to Saturn?

Assignments

Illustrate a page for Saturn in your notebook. Record all you have learned about this planet. Remember to use your best handwriting.

Make a Venn Diagram

Older Students: A Venn diagram is a wonderful tool that helps you compare and contrast different things. The drawing on the right is an example of what a Venn diagram looks like. Use a sheet of paper to make your own Venn diagram that you will use to compare and contrast Jupiter and Saturn. Study your notes from Jupiter (or read the Jupiter lesson again) to make sure you remember what you already learned about Jupiter. Write the things that are the same about both planets in the overlapping part of the diagram, and then record those things that are different in the ovals outside of the overlapping part. Place your completed diagram in your notebook.

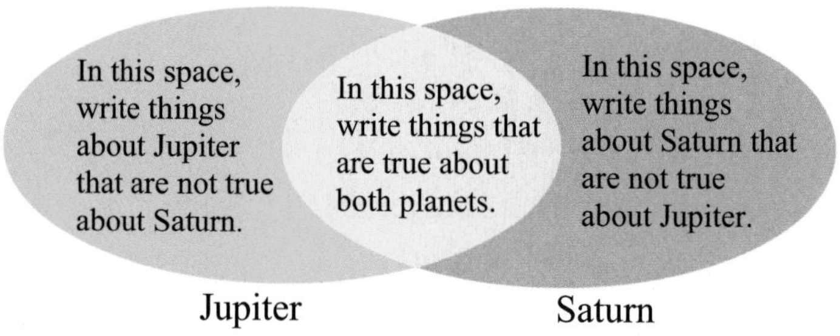

In this space, write things about Jupiter that are not true about Saturn.
In this space, write things that are true about both planets.
In this space, write things about Saturn that are not true about Jupiter.
Jupiter
Saturn

Younger Students: Copy the sample Venn diagram, and help your child fill it in so that she can compare and contrast Jupiter and Saturn. Place the completed Venn diagram in her notebook.

Project
Make a Centaur Rocket

Cassini was sent up into space with the help of a Centaur rocket weighing about 1,038 tons. These rockets are very important because they blast spacecraft, such as Cassini, out of our atmosphere and into space. Rockets launch into the air with a force that results from a chemical reaction. We are going to make our own rocket to explore how chemical reactions can launch a rocket.

This is the Centaur rocket that launched Cassini into space. Notice how small the engineers look!

Cassini was launched at night.

You will need:

- A plastic film canister (It must have a lid whose lip fits inside the canister rather than snapping on the outside of the canister. Film processing centers are happy to give them to you.)
- Eye protection, such as safety goggles or glasses
- Alka Seltzer® tablet
- Warm water
- Paper
- Tape
- Scissors

Follow the instructions on the next page.

Instructions

1. Tightly roll an entire piece of paper around your film canister. The canister should be upside down at the bottom of the roll. Make sure the lid can be easily removed from the canister.
2. Cut out a circle of paper large enough to be the top of the rocket.
3. Cut a section out of the circle as shown in the drawing on the right.
4. Form the circle into a cone and tape it to the top of the rocket.
5. Cut three triangles out of the paper and tape them to the bottom of the rocket (see the picture below).
6. Take the lid off the film canister.
7. Put on your eye protection.
8. Hold the rocket upside down, so that the film canister is now rightside up.
9. Fill the canister with warm water.
10. Go outside.
11. Make sure everyone is prepared to move away when you set the rocket on the ground.
12. Place an Alka-Seltzer® tablet in the canister.
13. Replace the lid quickly, and set the rocket down (rightside up) while you move away from it.
14. **MOVE BACK IMMEDIATLEY! IT LAUNCHES WITH A VERY POWERFUL FORCE!**

Cut this section out of the circle to form a cone.

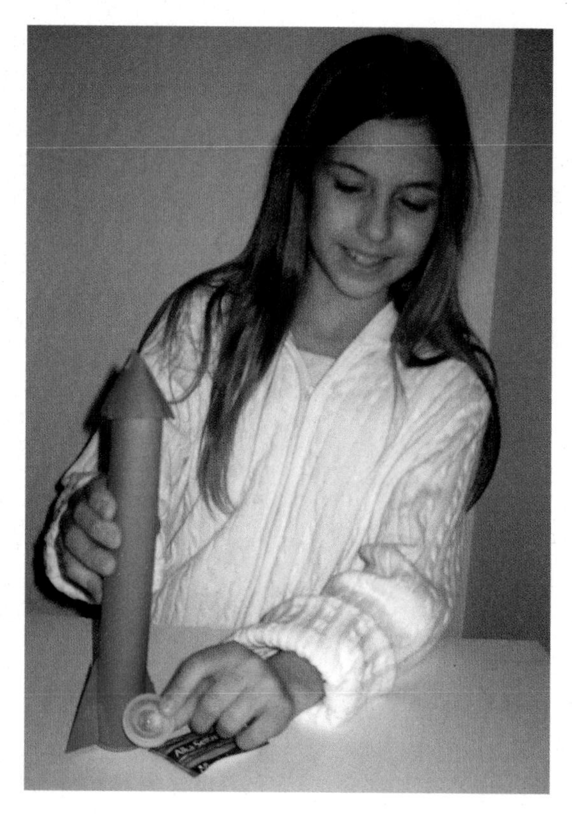

How does it work?

Your rocket launches because there is a chemical reaction when the Alka-Seltzer® mixes with the water. This reaction causes gases to build up in the film canister. These gases are very active, moving around a lot. As the reaction continues to happen, more and more gases develop. These gases need more room in which to move around, so they begin to push on the sides of the canister. They push harder and harder until they push the lid off the canister. When they do this, the gases fly out the open end of the canister, causing the whole canister to go flying into the air. Since the canister is a part of your rocket, it carries the rocket with it.

Lesson 11
Uranus and Neptune

Neptune

Two More Gas Giants

Beyond the two enormous yellowish-gold planets of Jupiter and Saturn are two enormous blue gas giants called Uranus (yur' uh nuhs) and Neptune (nep' toon). Uranus is blue-green, while Neptune is bright blue. Although Uranus and Neptune are the third- and fourth-largest planets in the solar system, they're so far from the sun that you need binoculars to see Uranus, and you usually need a telescope to see Neptune.

What makes these two planets appear blue and blue-green? It's their atmospheres, of course! Isn't that always the case with the gas giants? Uranus and Neptune both have helium and hydrogen in their atmospheres, like Saturn and Jupiter, but they also have another gas in their atmospheres: **methane** (meth' ayne). Methane absorbs red light and reflects blue light, so we see these planets as blue. Neptune is a deeper blue than Uranus because it has more methane in its atmosphere. Humans cannot breathe methane, so I guess we won't be vacationing on these two planets, either.

Uranus

We might call Uranus our sleepy planet, because it is lying down. The North and South Poles aren't up and down; they stick out the sides. God may have created Uranus as a tipped over planet, but it also might have just toppled over onto its side. Perhaps a giant comet hit it, or perhaps it was just supposed to be that way.

Because it is lying down, Uranus looks like a ball rolling around the sun rather than a top twirling around the sun. Most planets spin around like a top. To understand this better, do this activity: Place something to represent the sun on the table. Use a ball to represent Uranus, and roll it around the sun. Do you see how it is rolling? That is like Uranus. Now, to simulate the other planets, use another ball and hold it in place on the table. Begin to turn it as if you were unscrewing a lid. Now move it around the sun as you turn it like that. This is how most planets orbit. Do you see how Uranus behaves differently than the other planets?

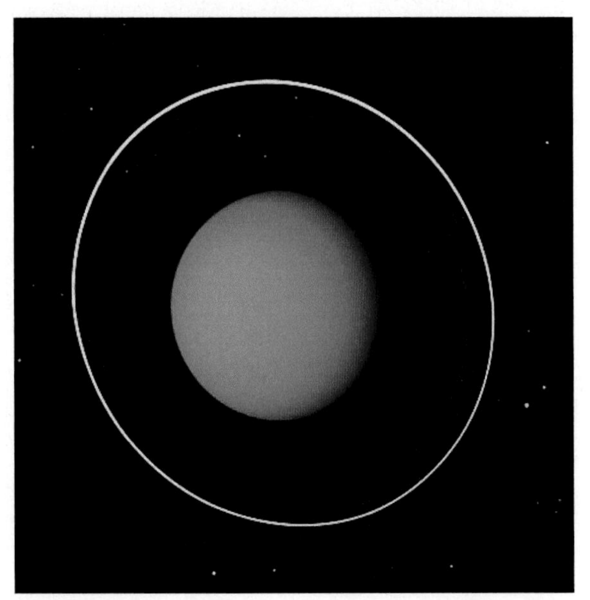

This is a picture of Uranus and its rings. Although it looks like there only are only six or seven rings, there are actually a total of eleven.

Uranus also has eleven rings that make its orbit seem even stranger. The rings are wound around Uranus vertically. That means they are up and down, rather than around the middle like the rings of Saturn.

The vertical rings and the rolling-ball effect make Uranus' rotation and orbit look like a wobbly wagon wheel that is about to fall off its hinges. Uranus is also spinning in the opposite direction as compared to most of the other planets. Can you think of another planet that spins in the "wrong" direction? It's Venus. You learned about its strange rotation in Lesson 4.

As you can imagine, Uranus is simply freezing cold. It's about 350 degrees below zero. That's 50 degrees colder than Saturn!

Moons

The rings of Uranus are filled with dust and rocks just as the other planets' rings are. It also has shepherd moons within the rings, holding the rings together: **Cordelia** (kor deal' yuh) and **Ophelia** (oh feel' yuh). These moons are named after characters in plays written by William Shakespeare.

Uranus used to be known as the planet with the most moons, until astronomers found even more moons orbiting both Jupiter and Saturn. Most of the names of planets and moons come from Roman myths and legends, but not the moons of Uranus. These moons are all named after characters in the writings of either William Shakespeare or Alexander Pope.

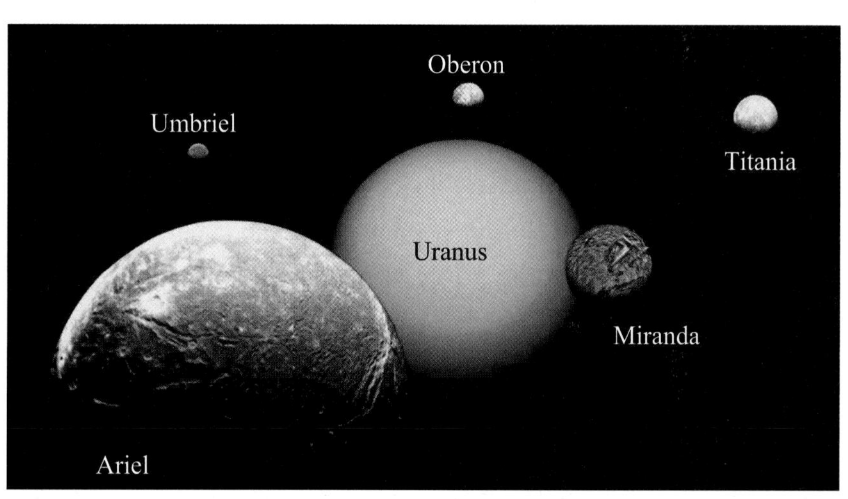

This is a composite photo of Uranus and its five largest moons. The sizes are not correct, because the artist wanted to make it look like you are closest to Ariel.

Eureka!

Eureka (yuh ree' kuh) means "I have found it." It is often used to mean an exciting discovery has been made. One such "eureka moment" was the discovery of Uranus in 1781. It was exciting because it was the first planet to be discovered in a long, long time! Mercury, Venus, Mars, Jupiter, and Saturn were well known throughout history. Uranus was something completely new!

William Herschel and his sister Caroline accidentally discovered Uranus. Their dad was a musician and taught them his favorite subjects: music, math, and astronomy. They lived and were homeschooled in Germany, but they grew up to be very important musicians in England. Before they discovered Uranus, William and Caroline made money singing and playing music in the opera. But, in his spare time, William built telescopes because he and Caroline studied astronomy. They were amateur (beginner) astronomers, just as you are!

Once William and Caroline discovered Uranus, they became very famous. They left their jobs in music and were paid by the government to be full-time astronomers. That was back when women were not usually given such important jobs. Caroline discovered many comets in her work as an astronomer, and William discovered many things, like two of Saturn's moons. Astronomers honor the work of the Herschels by naming the most interesting crater in the solar system the "Herschel crater."

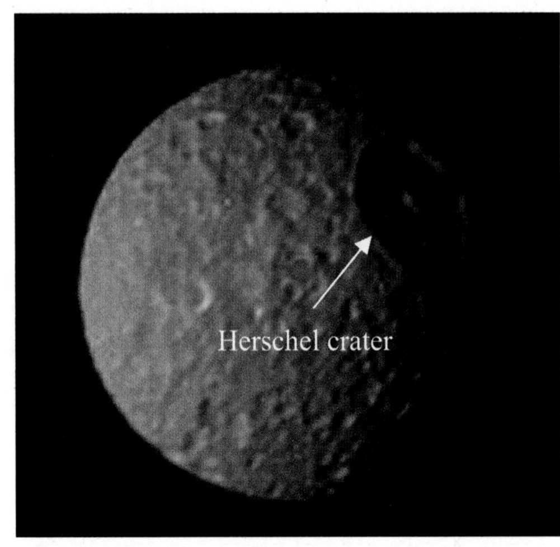

This is a photo of Mimas, a moon of Saturn that was discovered by William Herschel. The huge crater that makes this moon look like a Star Wars "death star" is called Herschel crater.

The discovery of Uranus happened one day as William and Caroline were looking through one of their homemade telescopes. They noticed a light that looked like a disk, not a star. Hundreds of years before, Galileo had discovered that planets look like disks. William and Caroline, however, were a bit obsessed with comets, and they thought it must be a comet. As they continued to watch this disk for months, they noticed that its orbit around the sun was nearly circular. This, of course, meant that it was a planet, not a comet! Remember, comets have very elliptical orbits, while planets have orbits that are nearly circular.

Guess what happens when you find a planet? You get to name it! Well, Herschel was not an ancient Roman who believed in false gods. In fact, he was a devout Christian. Because of this, he didn't want to name Uranus after a Roman god. Instead, he wanted to name Uranus after his king, King George III. Not everyone loved King George, however, so many people wanted Herschel to name Uranus after himself, calling it planet Herschel. In the end, they decided to keep with tradition and name Uranus after another Roman god. It is such a shame that the planets in God's wonderful creation, which were made for His glory, have been named after Satan's counterfeit gods. One day, when we are in heaven, we will call the planets by God-honoring names.

Orbit and Rotation

Because Uranus is so far away from the sun, it takes a long time, almost 84 years, for it to make one revolution around the sun. By the time a year passes on Uranus, your life will be almost over. Have you noticed that the farther a planet is from the sun, the longer it takes for it to go one time around the sun? Although each year on Uranus is extremely long, each day is just over 17 hours, which is not too different from our 24-hour day on earth.

What Do You Remember?

What chemical makes Uranus blue-green in appearance? Why does Uranus look like a ball rolling around the sun? What makes it look like a loose wagon wheel? Why was it so exciting to discover Uranus? Who discovered Uranus? How were they educated? How long does it take Uranus to orbit the sun?

Neptune

Neptune is the brightest blue planet in the solar system. It is a beautiful sight to behold. I'm sure that God must take great pleasure in this gorgeous creation. The Bible tells us that the highest heavens belong to God, but the earth was created for man. It must have given Him joy to create such amazingly beautiful spectacles in the heavens. They are His, and He made them lovely for Himself and the angels to enjoy. Of course, God also knew that we would one day be able to see His lovely outer planets and enjoy them as well. As the Bible says, "The heavens are telling of the glory of God; And their expanse is declaring the work of His hands." (Psalm 19:1)

Neptune's color is a beautiful, bright blue. The Great Dark Spot will be discussed later.

Eureka!

Upon finding Neptune, scientists might have once again yelled "Eureka," because they had been looking for it for quite some time! How did they know that Neptune was out there? Well, astronomers noticed that Uranus moved as if it was being pulled by another large object. They thought that this object must be a big planet, so they began looking for it. Unlike Uranus, then, Neptune wasn't found accidentally. Instead, its discovery was the result of a long search.

After finding Neptune, astronomers noticed that its orbit wobbles as well. At first, astronomers thought that this meant it was being pulled by yet another planet, so they looked and looked for one. They thought they found the culprit when they found Pluto, but they quickly realized that Pluto is too small to pull very strongly on Neptune. Astronomers resumed their search for this mystery planet, but very recent work indicates that the wobbles seen in Neptune's orbit are to be expected because of the gravity from the known planets in the solar system. As a result, most astronomers no longer think that there is another planet in the solar system.

Number Eight or Nine?

Neptune is sometimes the farthest planet from the sun! I know, I know, you may think Pluto is farther from the sun than Neptune. From 1979 to 1999, however, Neptune was the ninth planet from the sun, and Pluto was the eighth. In 1999, Neptune finally took its place as planet number eight, and Pluto hopped back to number nine. Why did this happen? Well, Pluto's orbit is more like a comet's orbit. In other words, it is more of an ellipse. Because of this, it crosses over Neptune's orbit for 20 years. During that time, it is closer to the sun than Neptune, and that makes Neptune the planet that is farthest from the sun.

Atmosphere

Neptune is a lot like Uranus, cold and covered with methane ice. You can see a striped pattern on Neptune, just like on Jupiter and Saturn. Can you guess what the stripes are? They are clouds. Just like Jupiter and Saturn, there are storms raging on Neptune. One of these storms can be seen in the picture of Neptune on the previous page. It is called the **Great Dark Spot**. It's just like Jupiter's Great Red Spot. It's as big as the whole earth and is filled with powerful winds. Interestingly enough, the Hubble Space Telescope has shown that the Great Dark Spot is now gone. This tells us that the storm is over. However, another large storm has formed another dark spot on a different part of Neptune.

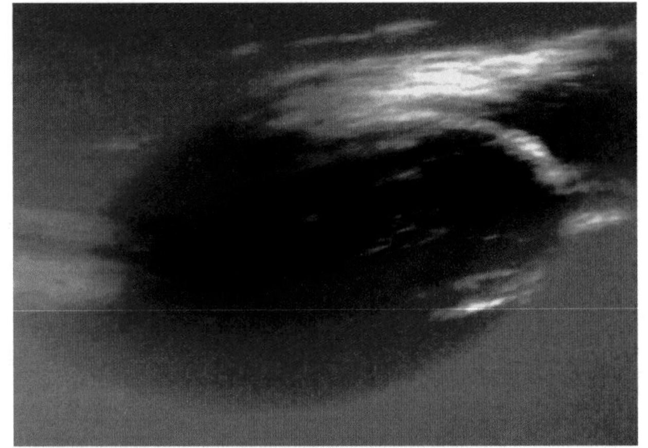

This is a closeup of Neptune's Great Dark Spot. Even though this storm is over, another one has appeared in Neptune's Northern Hemisphere.

When it storms here on earth, it rains, doesn't it? What is that rain made of? It is made of water. Well, some astronomers think that the rain in the storms of Neptune is actually made of diamonds! Yes, I said diamonds! Astronomers have done experiments that show that in the conditions they think exist in the storms on Neptune, methane can be turned into diamond dust. Since there is methane in Neptune's atmosphere, they think that diamond dust is formed in the storms and then falls from the clouds. In other words, they really think it is raining diamonds on Neptune!

Neptune is smaller than Uranus, so it's the smallest of the gas giants. It has a rocky core that astronomers believe has about the same mass as earth. That core is surrounded by liquid, which is then surrounded by the hydrogen, helium, and methane gases that make up its atmosphere. The winds on Neptune are even stronger than Saturn's winds. It would be a terribly blustery place to send a spacecraft.

Rotation and Revolution

Neptune is farther from the sun than Uranus, so you can be sure its year is longer than a year on Uranus. In fact, it takes about twice as long for Neptune to get around the sun: 164 earth years! Because it is so far from the sun, it is also cold. The temperature on Neptune is usually about 370 degrees below zero. Any spacecraft that will go to Neptune will have to be unmanned. Can you guess why? It would be just too cold to send a man to Neptune. A day on Neptune is about 16 hours, which is similar to Uranus's 17-hour day.

Moons

Astronomers have discovered more than ten moons around Neptune. The biggest is **Triton** (try' tuhn), and it's the coldest object that any spacecraft has ever visited. The average temperature on Triton is 400 degrees below zero! It revolves around Neptune opposite of the direction that Neptune rotates. It is the only moon known to do that. It is also moving closer and closer to Neptune each day. These two facts make most astronomers think that Triton was not originally Neptune's moon. Instead, it was probably captured by Neptune's gravity. Triton is a fascinating moon because it is filled with **geysers** (guy' zurs). Geysers on the earth are holes in the ground that spew out hot water from deep within the planet. You can visit some of earth's geysers at Yellowstone National Park. Triton's geysers are probably not spewing water. Instead, they are probably spewing a mixture of many other chemicals.

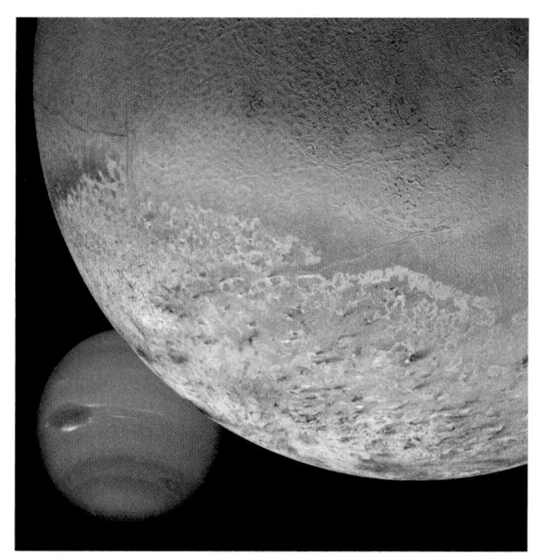

This is a photo of Neptune as seen from behind Triton. Triton is not bigger than Neptune; it just appears bigger because Triton is closer to you in this picture.

What Do You Remember?

Why was Neptune discovered? What made astronomers think there was another planet beyond Neptune? What chemical gives Neptune its blue color? Is Neptune the eighth planet from the sun? Explain your answer. How long does it take Neptune to revolve around the sun? What was the Great Dark Spot? What is the name of Neptune's biggest moon? What are geysers? Is water coming from the geysers on Triton? Remember to explain what you learned to someone else, so that you will lock the information into your brain.

Assignment

Begin with two separate sheets of paper. Title one "Uranus" and the other "Neptune." Make an illustration for each planet. Record all you remember about Uranus and Neptune under their titles. Don't neglect to color the planets with the beautiful blues that God gave them.

Activity

Create a Play about the Discovery of Uranus

You are going to create a short play about the discovery of Uranus. Older students will write it out; younger students will dictate their ideas and words to their parent / teacher to write out. Everyone should act it out! I will remind you of the facts mentioned in this chapter, and I will add a few more facts, in case you want to make your play even more interesting.

William and Caroline Herschel were brother and sister, living in Germany. Their parents had six children all together. Their father was a musician and taught all of his children music, as well as mathematics and astronomy.

Suggestion: You could do a scene with them at home learning music, astronomy, and math.

William and Caroline moved to England. This made their mother mad because she wanted Caroline to be her servant and housekeeper for the rest of her life.

Suggestion: You could do a scene with their mother begging them to stay so Caroline could be her servant.

In the new city, they became very important musicians in the opera. William played the piano and wrote the plays for the opera, and Caroline was a singer. In William's spare time, he built telescopes. He liked to make them more and more powerful to see deeper and deeper into space. People in the town loved and bought his great telescopes.

Suggestion: You could do a scene with them in the opera or making telescopes and selling them.

One day, William and Caroline noticed something in the sky that didn't look like a normal star. It looked like a disk. They thought it must be a comet. They watched it and watched it and debated about it. Finally, its nearly circular orbit made them realize it was a planet!

Suggestion: This is the most important scene to have in your play.

After this great discovery, the government paid them to be full-time astronomers and telescope builders. They discovered many more comets and other important things.

Suggestion: You can end with the government making them full-time astronomers.

Instructions

1. Decide how many people will be in your play. You may decide to have only William and Caroline. If you have lots of friends or family who would enjoy acting in your play with you, you could include William and Caroline's parents, brothers and sister, the government officials, other singers in the opera, and people coming to their door to buy telescopes.

2. Make a list of characters and their names.

3. Decide if this will be a short play with one or two scenes, or a longer play with up to five scenes.

4. Decide how the play will end:
   - Will it end with the discovery of Uranus?
   - Will it end with the government paying them?
   - Will it end with them being important astronomers, discovering many other things?

5. Decide where the first scene will take place and write what will happen in that scene.
   - Will it be at home in Germany with Dad teaching everyone math?
   - Will it be with them departing for England, thanking their dad for teaching them so well?
   - Will it be with them performing at the opera or building and selling telescopes?
   - Will you only have one scene in which they are looking through the telescope on the day they discover that Uranus is a planet?

6. Think about what people will say to one another in this scene and write it out. In plays, the name of the person speaking is written first with a colon after it, like this:

   Caroline: I think it's a comet.
   William: But it looks like a disk! Comets look like smudges.
   Caroline: What else could it be? Surely it is not another planet!
   William: Why, I never thought of that.

7. Repeat steps 5 and 6 for each scene.

This activity will be a lot of fun. It will be even more fun to act it out and use "props," such as telescopes. By doing this, you will never forget the very important story of the discovery of Uranus.

Project
Make Clouds

To understand why gas giants like Uranus and Neptune have clouds in their atmospheres, we will make clouds in a jar.

You will need:

- A glass jar
- Some ice
- A match
- A Ziploc® bag large enough to cover the opening of the jar
- Some very hot water

1. If the jar has a lid, remove it.
2. Turn on the hot water and let it run so that it gets really hot. To make this work even better, have your parent / teacher put water in a coffee mug and heat it up in a microwave oven.
3. As the water is heating up, fill the bag with ice and zip it closed.
4. Fill the jar halfway with the hot water. **Be careful! Don't burn yourself!**
5. Have your parent / teacher strike a match and drop it into the water so that it goes out when it hits the water.
6. Place the bag of ice over the top of the jar.
7. Watch as a cloud forms in the jar.

The cloud forms in the jar because the hot water evaporates, turning into a gas, which we call "water vapor." The water vapor rises, but it then cools when it hits the colder air that has been cooled down by the ice. This cooling causes the water vapor to become a liquid again, creating a cloud. The smoke from the match supplies some particles in the air above the water. This makes it easier for the water to turn into a liquid when it cools.

This is what happens on the gas giants. The inner core of each planet is very hot, while the gases that make up its atmosphere are extremely cold. Gases rise from the hot core, but when they get cooled off in the icy-cold atmosphere, they turn into liquids and then solids (like ice crystals), making clouds. Now remember, the clouds on the gas giants are not made of water. They are made of methane and other chemicals. Still, the overall process is very similar to what you saw here.

Lesson 12
Pluto and the Kuiper Belt

This is an artist's idea of what it might look like to see a Kuiper belt object in the foreground with Pluto behind it. This is not a photograph.

Kuiper Belt

You may have been hoping to hear about Pluto next, but objects in the **Kuiper** (ky' pur) **belt** are sometimes closer to the sun than Pluto. Because of this, we will take a quick glance at the Kuiper belt before we study Pluto. The Kuiper belt is a lot like the asteroid belt, but the objects in it are much larger and much colder. They look a lot like comet nuclei (new' klee eye). "Nuclei" is the plural of nucleus. Because Kuiper belt objects look like comet nuclei, many astronomers think that the Kuiper belt is the source of the solar system's short-period comets. As I mentioned in Lesson 8, however, there are not nearly enough objects in the Kuiper belt to explain all of the short-period comets that we see today. If the Kuiper belt is a source of short-period comets, then, it is not the only source.

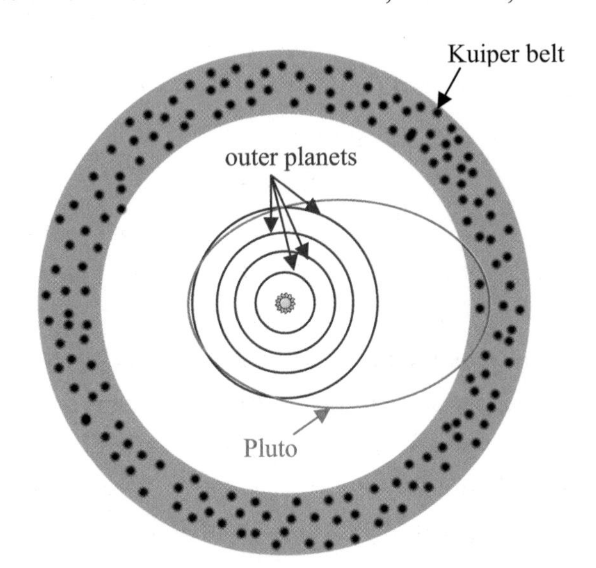

This is a drawing of the orbits of the outer planets (blue circles), Pluto (red ellipse), and the Kuiper belt. The drawing is made as if you are above the solar system, looking down on it.

The objects in the Kuiper belt look a lot like Pluto. Many of them have been labeled **Plutinos** (ploo teen' ohs), or little Plutos. Because these objects are so similar to Pluto, some astronomers think that Pluto was once a part of the Kuiper belt. They think that a collision between Pluto and another Kuiper belt object knocked Pluto out of the Kuiper belt and into an elliptical orbit around the sun. It is also thought that Neptune's moon, Triton, as well as at least one moon of Saturn are Kuiper belt objects that got trapped into orbit around those planets.

The Kuiper belt is named after Gerard Kuiper, the man who first guessed that it might exist. He had never seen the Kuiper belt, but he thought that there must be some space rocks orbiting the sun beyond the orbit of Neptune. This was just an idea for over 40 years, but then in 1992, a large object was discovered in the part of the solar system that Kuiper had said the belt should be. This object, named 1992QB1, was about 150 miles wide, and it looked like a smaller version of Pluto. Several similar objects were quickly discovered, and the Kuiper belt went from being just an idea to being a definite part of the solar system. Since then, the Kuiper belt has been studied by many astronomers, and many more Kuiper belt objects have been discovered. In 2002, for example, an object that is about half the size of Pluto was found in the Kuiper belt. It is called **Quaoar** (kwa' whar).

This picture allows you to compare the sizes of the earth, earth's moon, Pluto, and Quaoar.

Pluto

It seems that many people's favorite planet is Pluto. Maybe that's because it's so small. Perhaps people like it because it is so far away. Maybe it is a favorite because the name is catchy and reminds us of a cartoon character.

Do you remember how Pluto was discovered? Astronomers were looking for a planet that was pulling on Neptune, making Neptune sway back and forth a bit. They just knew there was another planet out there, and while they were looking for that planet, they discovered Pluto! Thinking Pluto might be the planet pulling on Neptune, they named Pluto as the ninth planet. As they began to really study Pluto, however, they realized that this tiny planet couldn't be pulling on Neptune! Pluto's mass is so small that its gravity is very weak. The gravitational pull that Pluto exerts on Neptune is simply not large enough to affect Neptune's orbit. Of course, we now know that Neptune's orbit can be explained with the planets that have already been discovered, so most astronomers don't think there is another planet to be discovered.

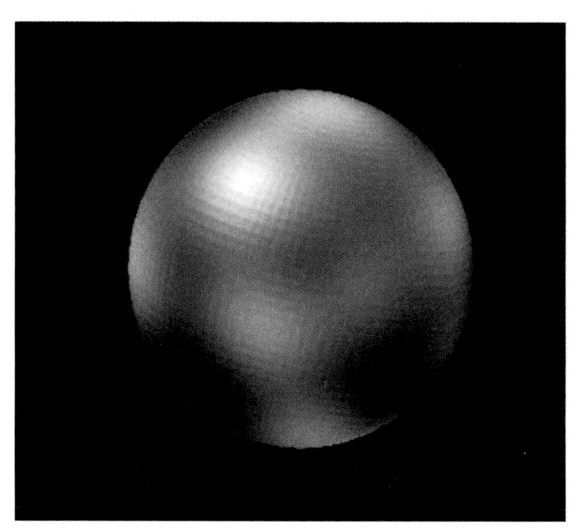

Pluto has never been visited by a spacecraft. As a result, this is probably the best image we currently have. It was taken by the Hubble Space Telescope.

Who's Number Nine?

In 1979 the solar system became a bit mixed up. That's when Pluto moved closer to the sun than Neptune. Every 248 years the two planets swap places for about 20 years. During this time, Pluto becomes the eighth planet and Neptune the ninth. For part of my life, then, Pluto has been the eighth planet, not the ninth planet, from the sun! In February of 1999, Pluto became the ninth planet again as its 20-year trip inside the orbit of Neptune was over.

Can you remember all that you have learned so far about the Kuiper belt and Pluto? Explain Pluto and the Kuiper belt to someone else in your own words.

Hubble Space Telescope

Pluto was discovered in 1930, but it was very hard to study with the telescopes we had back then. We can see it so much better now that we have the Hubble Space Telescope. Do you remember that Hubble is a giant telescope floating up in space?

Studies using the Hubble Space Telescope have shown us that Pluto is probably a big block of ice with some rocks mixed in. Does that sound familiar to you? It should. That's what a comet's nucleus is made of. In other words, Pluto looks a lot like a comet's nucleus. As you might expect, it is also very cold on Pluto. As far as we can tell, Pluto can get as cold as about 390 degrees below zero! We know that Pluto does have an atmosphere, but since we have not sent a spacecraft to Pluto, we don't know much about it. We know that it is very thin, and we think that it has methane and a few other gases in it. Even though its atmosphere has methane, Pluto is not blue, because there just isn't enough atmosphere to give Pluto a distinct color.

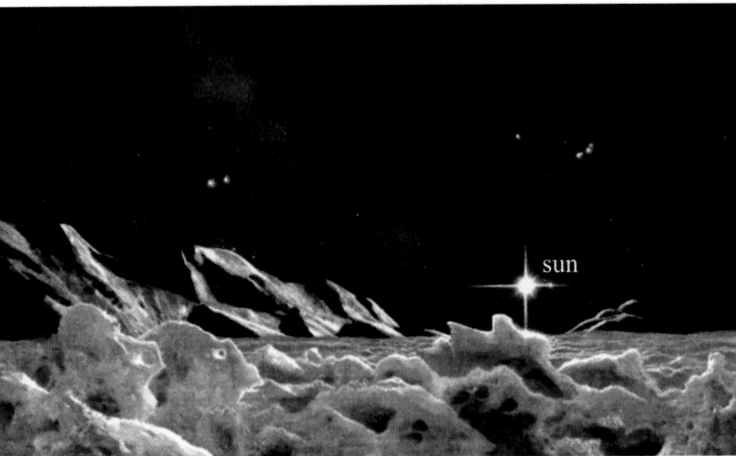

This is an artist's idea of what it might look like to see the sun from the surface of Pluto. Notice the ice on the surface, and notice how small the sun looks. That's because it's so far away.

Opposite Rotation

Pluto rotates in the opposite direction as compared to the earth and most of the other planets. Like Uranus and Venus, it is spinning the wrong way! It spins slowly compared to earth, because it takes just over six earth days for a day to pass on Pluto. This little planet also has something else in common with Uranus. Do you remember why we called Uranus the "sleepy planet?" Well, Pluto must be sleepy as well, because it lies on it side, just like Uranus.

Winter's Coming

Pluto hasn't changed much since we first started studying it in 1930. Of course, that's to be expected, since it takes *248 earth years* for Pluto to make just one trip around the sun. Since the time we have been studying it, then, it has traveled just over one quarter of the way around the sun. During this time, it has been fall on Pluto. Winter is coming, however, and astronomers are hoping to see what kinds of changes will occur. As far as they can tell so far, Pluto seems to be warming up just a little, even though astronomers expected it to cool down. This is strange, because the distance between Pluto and the sun is also increasing. Between the fact that winter is coming and the fact that Pluto's orbit is putting it farther from the sun, astronomers thought that they would see Pluto cooling down. So far, that's not what has been observed.

Obviously, we have a *lot* to learn about Pluto. That's why NASA would like to send an unmanned spacecraft to Pluto to study it. The problem is, such a mission would be very expensive, and NASA has only so much money to spend. Currently, NASA is planning to build a spacecraft called **New Horizons**. If things go as planned, it should launch from earth in January of 2006. If it is successful, it should reach Pluto by 2015. That should give you an idea of how far away Pluto is. It will take a spacecraft almost 10 years to travel there! After New Horizons studies Pluto for a while, it will continue into the Kuiper belt to study the icy bodies that are found there.

This is an artist's idea of what it will look like when New Horizons is studying Pluto.

Moon

Pluto has a moon. Yes, this tiny little planet has its own little moon, called **Charon** (kair' un). Charon is really small, but not that much smaller than Pluto. It took us a long time to find this moon. Remember, Pluto was discovered in 1930. It took 48 more years for us to find Charon. Why did it take so long? Well, part of the problem is that Charon is small, like Pluto. The other problem is that Charon is very close to Pluto. This makes it hard for astronomers to see it in a telescope.

This is an artist's idea of what Pluto (the larger object) and Charon (the smaller one) look like.

Some astronomers think Pluto's moon is just too big to be a moon. After all, no other moon in the solar system is that big compared to the planet that it orbits. Please understand what I am saying here. Charon is very small. However, compared to Pluto, it is really much bigger than it should be. To give you an idea of what I mean, suppose you had a bunch of balls the size of earth's moon. You could fit about 50 of those balls into the earth. On the other hand, you could only fit about 6 Charon-sized balls into Pluto. This is what I mean when I say that compared to Pluto, Charon is very large for a moon. Because of this, some astronomers think that we shouldn't call Charon a moon at all. Instead, we should call it another little planet stuck in orbit with Pluto. In fact, some astronomers think that rather than calling Pluto a planet, we should call both Pluto and Charon a "double planet."

What Is Pluto Anyway?

The more we learn about Pluto, the more it seems that Pluto isn't the planet we thought it was. There has been a big debate among astronomers about whether Pluto can even be called a planet. Ever since astronomers found the Kuiper belt, they have been wondering whether Pluto is really a planet at all.

Is Pluto a planet or a comet? Let's explore the evidence on both sides of the issue. We will start with the complaints from the **"Against"** side, and then we will look at the theories of those **"For"** Pluto being a planet. After we discuss all the details, I will give you the opportunity to decide what you believe about Pluto's status as a planet.

Planet Pluto or Comet Pluto?

Pluto Is the Size of a Moon

Against: Pluto is simply too tiny. It's the smallest planet in our whole solar system. Pluto is about the size of the United States of America. It is actually smaller than many of the moons in our solar system. Pluto is smaller than Earth's moon, Io, Europa, Ganymede, Callisto, Titan, and Triton.

For: Mercury is smaller than Jupiter's largest moon, Ganymede. We know for certain that Mercury is a planet. This means that being smaller than a moon doesn't disqualify an object from being a planet.

Off Center Orbit

Against: Pluto's orbit is the most highly elliptical of all the planets. It is also the only planet that comes inside the orbit of another planet, the way comets do.

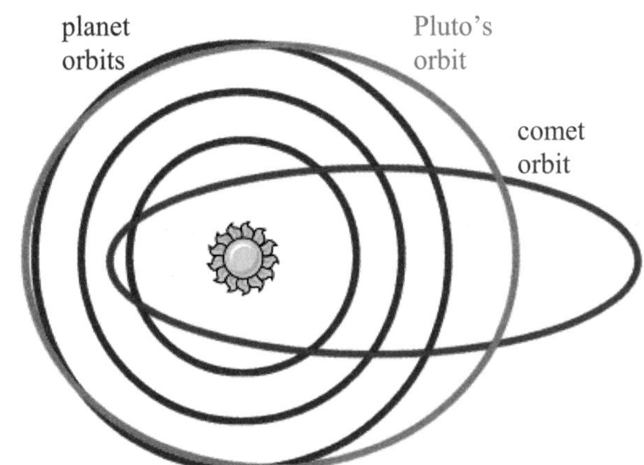

For: Crashing into a comet, like Hale-Bopp, could have made the orbit of Pluto different from the other planets. At one time, Pluto's orbit might have been similar to the other planets'. The orbit is also not nearly as elliptical as a standard comet. In fact, it is only a bit more elliptical than Mercury's orbit. Given these facts, Pluto's orbit still looks more like that of a planet than that of a comet.

Inclining Orbit

Against: Most of the planets orbit around the sun on close to a flat plane. Pluto's orbit slants up and down. Most comet orbits slant as well. If Pluto were, in fact, a comet, then it would be natural for it to orbit at a slant.

For: A crash with another large object could have altered its orbit to make it slanted.

Pluto's Moon Is Too Big

Against: Pluto is the only planet that has a moon that is so big compared to the size of the planet. The other planets' moons are small compared to the planets they orbit. Pluto and its moon could really be comets that got caught orbiting together. Many scientists consider Pluto and Charon a double planet system, rather than a planet with a moon.

This is an image of Pluto and Charon as seen from the Hubble Space Telescope.

For: Charon isn't evidence against Pluto being a planet, because Charon could have been a comet that got pulled into orbit with Pluto, just as Triton is most likely a comet that got pulled into orbit around Neptune.

Misfit with Gas Giants

Against: Pluto is a misfit. The planets beyond the asteroid belt are gas giants.

For: There is no reason to think that all of the planets past the asteroid belt should be gas giants. The coldest outer reaches of the solar system should be home to planets made of ice and rock.

Looks Like a Comet

Against: Comets are big blocks of ice with rocks and rubble mixed in. They do not have a hot inner core as all the other planets do. Pluto is the only planet that does not have a hot inner core.

For: All the comets we are familiar with make many changes, forming comas and burning off material. Pluto doesn't have those kinds of changes.

Against: The reason comets burn off material and undergo change is because their orbit brings them close to the sun. Pluto never comes near the sun. Comet-like objects that do not come near the sun do not change.

For: The comet-like objects that do not come near the sun are in the Kuiper belt. Pluto is not in the Kuiper belt.

Looks Like a Kuiper Belt Object

Against: You can't tell Kuiper belt objects apart from Pluto. Except for size, Pluto is nearly identical to the objects in the Kuiper belt. Many Kuiper-belt bodies have been labeled Plutinos, which means little Plutos. If the Kuiper belt and Quaoar had been discovered when Pluto was, Pluto almost certainly would not have been classified as a planet in the first place.

For: Pluto is a lot bigger than all the known objects in the Kuiper belt. Therefore, it should not be classified as Kuiper belt object.

What Is a Planet?

For: The best way to decide whether Pluto is a planet is to ask the question, "What is a planet?" A planet could be considered any large body that orbits the sun, having a strong enough gravitational pull to stay in orbit.

Against: Depending on what you mean by "large," many objects in space, including comets, could be called planets.

Now You Decide

The evidence is in. It's time for you to make your own hypothesis (hi pahth' uh sis). "Hypothesis" means "a good guess based on the facts." At the end of this lesson, you can report on your decision in your notebook.

What Do You Remember?

What is the Kuiper belt? How was Pluto discovered? What are some of the strange features of Pluto? Why do some astronomers believe Pluto is not a planet? What do these astronomers think it is?

Assignment

Illustrate a picture of Pluto and the Kuiper belt. Record all the facts that you can remember about Pluto and the Kuiper belt. Place your work in your notebook.

Older Students: Which side of the Pluto debate are you on? Write out your argument for or against Pluto as a planet. On another sheet of paper, create the point system illustrated below. In the center circle, write your hypothesis. In other words, write what you think about Pluto. You could write, "I do not believe Pluto is a planet." You could write, "I believe Pluto is a planet." In the smaller circles, write one or two words that describe each point you believe makes your hypothesis true. You can look back at the lesson to remind yourself of the different points. If you must use more than two words, that is okay. Try not to use sentences. In the lines next to the circle, give a few sentences to support your point. You don't have to use all five circles if you do not have five points to support your hypothesis.

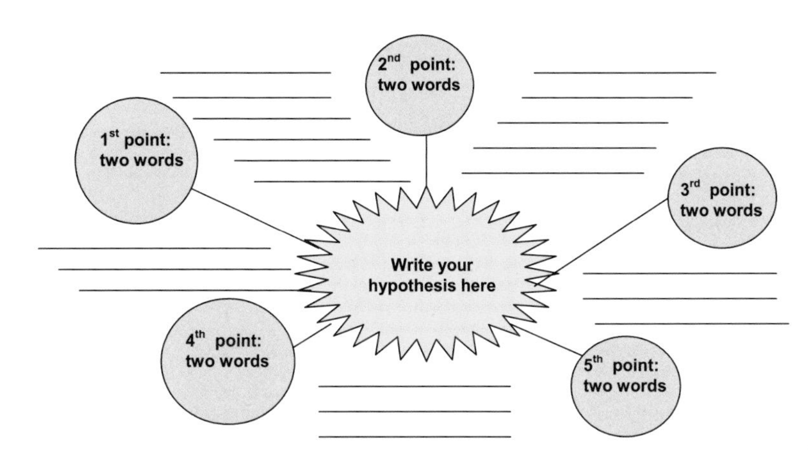

Make a Book

You are now going to make a very interesting book about your hypothesis. You will use all of the points you just made. This will definitely be a book people will want to read, so do a great job!

Get several pieces of paper that you can staple or bind together. Place your title on the cover. Write out your hypothesis on the first page. On each new page, write out a sentence or two for each point you have that supports your hypothesis. Make a cover page illustration and illustrate the inside pages of your book, as well. Staple or bind the pages together, and you have a book!

Younger Students: You have now completed the study of all the planets in our Solar System! Because you have learned so much, you're now qualified to make your own solar system book. You will need six pieces of paper, folded in half and stapled in the middle. Begin by drawing a different planet on each page. Take care to do your best work; you will want to share your solar system book with your family and friends. After you have illustrated each planet, write or dictate to your parent / teacher some interesting facts about each planet. You may want to look through your notebook to remind yourself of the things you have learned. When you come to Pluto, you can include what you believe about whether Pluto is a planet or a comet and why. Make an interesting illustration for your front cover. Now you have a book about the solar system!

Project
Make Ice Cream!

Pluto is a very cold planet, and everything on it is frozen. To freeze something, you must cool it down. How much do you have to cool it down? That depends. Different chemicals have different freezing points. For example, water freezes at 32 degrees, so we call 32 degrees the **freezing point** of water. Other chemicals have other freezing points. If you mix other chemicals with water, the freezing point of the mixture is lower than the freezing point of water. We are going to do an experiment to see how this works. The amounts listed make one small serving. You can increase the amounts to make more.

You will need:
- 2 tablespoon powdered sugar
- ½ cup whipping cream (Whole milk or half-and-half will work.)
- ¼ teaspoon vanilla
- 6 tablespoons rock salt
- 1 pint-size Ziploc® plastic bag (you need a bag that will not leak or break open)
- 1 gallon-size Ziploc® plastic bag
- Several ice cubes

Instructions

1. Put the milk, vanilla, and sugar into the small bag, and seal it.
2. Fill the large bag half full of ice, and add the rock salt.
3. Place the small bag inside the large bag, and seal the large bag.
4. Shake vigorously until the mixture in the small bag is ice cream (about 6-8 minutes)
5. Wipe the salty water and ice off the top of small bag.
6. Open the bag and pour its contents into a bowl. Enjoy the results of your experiment!

How Did That Work?

Notice that the ice cream you made was solid. That's because it froze. However, the salt water in the outer bag was not frozen. Why did the mixture in the inner bag freeze when the mixture in the outer bag did not? Well, when the ice and salt mixed, the salt did two things. First, it lowered the freezing point of water, which made the ice melt. Second, it actually cooled off the melting ice, making the mixture colder. Usually, if water gets to a temperature of 32 degrees or below, it freezes. However, the salt not only cooled the ice, but it lowered the freezing point of the water, so the water got colder and still did not freeze! The mixture in the outer bag was actually about 28 degrees once the salt and water mixed well. This was cold enough to freeze the cream in the smaller bag, making ice cream! Complex things like this happen on planets, when lots of chemical mix together.

Lesson 13
Stars and Galaxies

Star Light, Star Bright

"He counts the number of the stars; He gives names to all of them." Psalm 147: 4

The night sky is peppered with many tiny stars, some dim and some bright. Some of the stars we see in the sky are actually planets, but without a telescope, they look like stars. These stars surround the earth all of the time, day and night, but our sun's light is so bright that it drowns out the light from the stars during the daytime. We can only see the stars when the earth is facing away from the sun. That way, the sun's light does not drown them out. The stars are farther away than the planets. They are outside of our solar system.

What is a star? A star is an object that is something like our sun. It is a large body that produces a lot of energy. It produces so much energy that we can see its light, even though it is far, far away. There are many billions of stars in the universe. New ones are discovered each day. Some have been named by man, billions more have not. Nevertheless, God said he calls each star by name. Each of the billions of stars in the sky is important to God. That's why He has a name for each one. That isn't a surprise, since there are billions of people on the earth. Each one of us has a name, and each one of us is on God's mind continually.

This is a photo of stars in the sky. The brightest star in the photo is called Deneb.

On a cloudless night, when the moon is new (meaning it is completely dark), you can see about 3,000 stars in the sky. With a fairly inexpensive telescope, you can see about 100,000. As we gaze at the stars, they appear to twinkle. That's because we have an atmosphere which surrounds the earth. The particles in the atmosphere are busily moving around. Every few seconds, these particles pass between a star and our eye. When this happens, the star dims for a quick second as the particle passes in front of it. The star then brightens again when the particle moves away. This makes it look like the star is twinkling. The star is not really twinkling, however. It is just the particles moving about in our atmosphere. The man-made satellites orbiting our earth (like the Hubble Space Telescope) take pictures of the stars. They are outside of our atmosphere, so they get very clear pictures. Our pictures here on earth are not as clear because of our atmosphere.

Different stars are visible in the sky each season as we revolve around the sun. This means we get a different view of the vast universe every few months! One of the most important stars, besides our sun of course, is the **North Star**, which is named **Polaris** (poh lair' us). The North Pole is always pointing toward Polaris, so even as all the other stars we see change, those who live in the Northern Hemisphere can always see the North Star. Also, since we know that the North Pole points towards Polaris, if you can see Polaris, you know which way is north. Because of this, Polaris was an important star for navigation back when people used the stars to figure out where they were and where they needed to go.

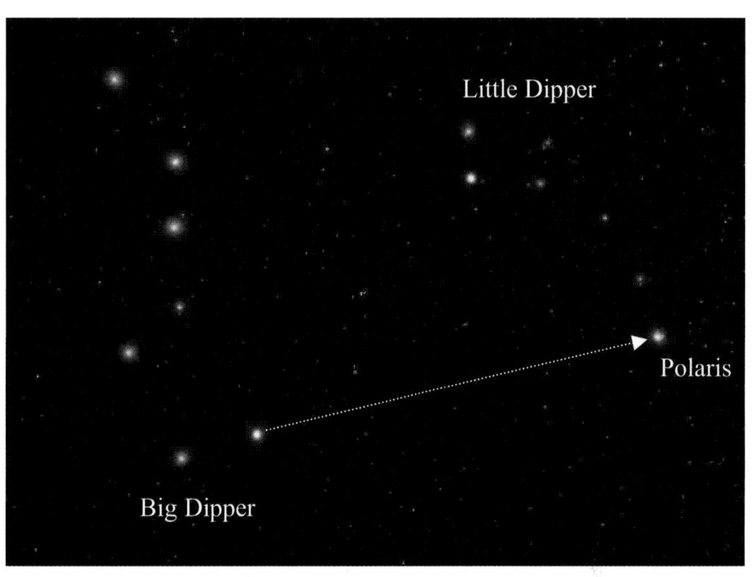

This is a photo of the Big Dipper and the Little Dipper. The "ladle" of the Big Dipper points to the North Star, which is on the "handle" of the Little Dipper. Please note that a photographic technique was used to make the asterisms stand out from the background. In real life, the asterisms are not this bright.

Polaris is part of a group of stars called the **Little Dipper**. Although many people call the Little Dipper a **constellation** (kahn' stuh lay' shun), it is not. It is actually called an **asterism** (as' tuh riz' uhm), which means it is a set of stars that form a shape that is easy to recognize. As you see in the photo above, Polaris and some other stars form what looks to be a spoon with a curved handle. That's the Little Dipper. It is called the Little Dipper because another set of stars form a bigger spoon shape. Not surprisingly, we call it the **Big Dipper**. One neat thing is that the two stars which make up the edge of the Big Dipper's ladle form a line that points to Polaris. When you learn about the constellations later on in this lesson, you will learn that the Big Dipper is a part of the constellation known as Ursa (ur' suh) Major, and the Little Dipper is a part of the constellation known as Ursa Minor.

All the stars appear to rotate around Polaris. To understand this concept, use an umbrella and a piece of chalk (or a paintbrush and glow-in-the-dark paint). Open your umbrella and look inside. Do you see that there are lines that make different sections of the umbrella? Those lines are a little like the segments that astronomers use to divide up the night sky for each month. Every month represents a different segment. The very tiptop where the pole

comes through the umbrella is where Polaris is located. Take your chalk (or paint) and make a mark around the center to make Polaris. Next, use your chalk (or paint) to make star patterns around the inside of the umbrella. You can spin the umbrella around and watch all the stars move around Polaris, the center. Focus your eyes on one segment of the umbrella. Within that segment are the stars you are able to see from earth that month. When you turn your umbrella, different stars come into view. The North Star, Polaris, remains in the same spot! If you look up at Polaris as you turn it, it looks as if all the stars are turning around Polaris. That is what it is like to view the stars from the Northern Hemisphere of the earth. The stars seem to rotate around Polaris.

The brightest star in the sky is not the North Star. It is **Sirius** (seer' ee us). Sirius is mentioned and described in ancient writings from 1,500 years ago. Sirius, like many stars in the sky, is actually a **double star system**, which is also called a **binary star system**. That means that there is another star in the solar system of Sirius that orbits around Sirius. The star that orbits around Sirius is dim and small. We call it a "white dwarf" star.

This is a drawing of a binary star system, where one star orbits another.

White dwarf stars are actually stars that are in the process of "dying." I bet you're saying, "Stars die?" Yes, they do. Stars have a limited lifetime, because the fuel that they use for thermonuclear (thur' moh new' klee ur) fusion eventually runs out. Do you remember what thermonuclear fusion is? You learned about it in Lesson 2. Well, thermonuclear fusion takes fuel. When that fuel runs out, the star begins to die, actually falling in on itself and becoming smaller. For certain stars, this can actually form a **black hole**. The Bible says that this world is passing away (Matt 24:35). Star deaths point to a decaying universe, just like the Bible says.

Black Holes

When a star runs out of fuel, it begins to collapse inward. This makes it get smaller and smaller. What makes a dying star collapse like that? The star's own gravity presses it inward. When this happens, the star begins to shrink. This heats up the core of the star. If the star has a lot of mass, the heat generated by the collapse is truly awesome, and the star actually explodes! This is called a **supernova** (soo' per noh' vuh). Lots of the star gets shot into space, but a lot of the core stays behind. At that point, the core is very small and has a *lot* of mass. Do you remember what mass does? It causes gravitational pull. Well, the core that is left behind is so small and has so much mass that its gravity starts pulling everything around (even light) into it. At that point, the object is dark, because light cannot escape it, and we call it a **black hole**. Strange, huh?

Scientists have never really seen a black hole directly, but there is a lot of evidence that they exist. We can't see them because they are black, and to see something with a telescope, it has to have light coming from it. However, the strong gravity in a black hole does affect the things around the black hole, and we can often see stars or gases behaving in a way that indicates there is a black hole near them. For example, we can see gases in space that orbit very quickly in tight circles. The speed at which they are moving, and the size of the circle they are moving in, indicates that they are orbiting a black hole. Black holes point to a universe that is groaning from the decay forced upon it by Satan and the sin of man.

This is an artist's idea of what it looks like when gases are spinning around a black hole. You cannot see the black hole; it is at the very center of the swirl. However, you can see all of the activity going on around it, and from the activity, you can conclude that the center is, indeed, a black hole.

Don't worry about us ever being sucked into a black hole, because we are getting farther and farther away from any black hole that may be out there. That's because the universe is expanding (getting bigger and bigger), sort of like a balloon gets bigger and bigger as you blow it up. That is what is happening to the universe. All the stars are moving farther and farther apart from one another. We are getting farther away from every star, black hole, and solar system that surrounds us.

To understand this better, you will need a balloon, a pen or marker, and a device for measuring, such as a ruler or tape measure. First, make little stars all over your balloon with the pen or marker. Measure how far apart the stars are from one another. Next, blow up the balloon and tie it closed. Measure the distance between the stars again. When the balloon expanded, the stars became farther apart, didn't they? We are getting farther away from the stars in the sky just as the stars on your balloon got farther away from each other. That gives you an idea of how our universe is expanding.

Supernovas

When I told you how a black hole forms, I mentioned that the collapse of some stars can lead to giant explosions, called supernovas. Believe it or not, we have actually seen this happen! In February of 1987, for example, three separate astronomers were looking at a portion of the sky and saw a very bright star that should not have been there. When they compared what they saw to images of the same part of the sky taken previously, they saw that, indeed, no bright star had been seen there before. The bright star remained there for several months and slowly dimmed away. It was quickly determined that this "star" was actually a supernova, and it was named supernova 1987a. Supernovas have been seen throughout history. One of the earliest recorded ones was seen by Chinese astronomers in 1054. That supernova was so bright that it was visible in the daytime!

What is really interesting about supernovas is that they leave behind a "footprint." The gases and dust that are thrown from the exploding star form a cloud that spreads out. This cloud of dust and debris is called a **nebula** (neb' yuh luh). Nebulae (the plural of nebula) have all sorts of interesting shapes and colors, depending on the type of star that went supernova. An example of a nebula is shown on the right. The Chinese astronomers who observed the supernova in 1054 recorded where they had observed it in the sky. When astronomers later pointed a telescope in that direction, they found this nebula, which is called the **crab nebula**. Thanks to the detailed records of Chinese astronomers from nearly 1,000 years ago, we now can see the nebula that is the "footprint" of the supernova that they saw.

This is a photograph of the crab nebula, which contains the remains of the first recorded supernova.

Variable Stars

Our sun is a special star, unlike many stars in the sky, because it always puts out the same amount of heat, light, and energy every single day. Over a long period of time, the heat, light, and energy is actually increasing (as you learned in Lesson 2), but on a day-to-day basis, the energy that comes from the sun is pretty much the same. We never wake up in the morning wondering if the sun is going to be so hot that all the water will dry up in the ocean today. We never go to bed worrying that the sun might become so dim that all that the whole world would freeze into an ice cube while we sleep. Believe it or not, astronomers have discovered that this is exactly what some other stars do. Sometimes they burn very hot and bright, and at other times, they do not burn very brightly, putting out less heat and energy. These stars are called **variable stars**, because the energy in them varies.

Do you remember what a binary (or double) star system is? It is a system in which one star orbits another star. Well, nearly half of all of the stars you see in the sky are double stars! Given all the double stars in the sky and all the variable stars in the sky, you can now see how special our own star is! In fact, most stars in the universe have many features that are different from our star, the sun. These features would make it impossible for us to live on a planet that orbited them. We could not survive if our special sun was not just the way God made it. Aren't you glad that God loves us enough to give us the perfect star to orbit?

Categorizing Stars

Hot or Cold

Scientists can tell how hot a star is by its color. Have you ever watched a fire burn in a fireplace or at a campsite? The coals at the bottom of the fire first glow red but then become white. The white coals are hotter than the red coals. If you want to roast your marshmallow quickly, you will use the white coals, because the color tells you that they are the hottest ones. Well, it turns out that the color of a star also tells you how hot it is. Blue stars are the hottest stars, while red stars are the coolest. Yellow stars (like our sun) have a medium temperature.

Sometimes, when people are dividing up children, they divide them up based on the first letter of their last name. What is your last name? If your last name is Smith, you would be in the S category. If we put people into categories like that, we would be arranging them by letter. That's what scientists do with stars. Each star is put in a letter category based on how hot it is. There are seven letters for stars: O, B, A, F, G, K, and M. The letter "O" stands for the hottest stars (the blue ones), and the letter "M" stands for the coolest stars (the red ones). Our sun is a G star, because it is a medium-temperature star.

The way astronomers remember these letters is by creating a mnemonic phrase. Do you remember doing that in the first lesson to learn the order of the planets? Astronomers make a lot of mnemonic phrases. One phrase they use to remember the order of the letters is, "Of Berkeley Astronomers, Few Give Kind Marks." Berkeley is a very tough college, and "marks" is another way of saying "grades." So this sentence is easy to remember for anyone who studied astronomy at Berkeley! Do you see the first letters of the words in that sentence? They are O, B, A, F, G, K, and M. That's the order of star categories from hottest to coolest. Later on, you will make your own mnemonic phrase to help you remember the star categories.

Bright or Dim

What grade are you in? Most grades are assigned a number aren't they? If you are in first grade, you are classified under the number "1." If you are in fourth grade, you are classified under the number "4." If your last name is Smith and you are in the fourth grade, you could be put in the category "S-4." That would tell someone that your last name starts with "S" and that you are in the fourth grade. Stars are also assigned a number as well as a letter. The number tells you how bright or dim the star is. Oddly enough, a smaller number means a brighter star. If a star is given a "1," for example, it is much brighter than a star that is given a "9." When you put the letter and the number together, you learn a lot about a star. Our sun, for example, is a G-5. The "G" tells us that it is a medium-temperature star, and the "5" tells us that it is dimmer than Sirius, which is an A-1 star. Do you remember what Sirius is? It is the brightest star we see in the night sky. Wait a minute. How can the sun be dimmer than Sirius? The sun is so bright that it drowns out all of the stars, including Sirius. Well, when astronomers talk about how bright or dim a star is, they do not mean how bright or dim it *looks from earth.* They mean how bright or dim it would be *if you were close to it.* The sun looks very bright to us because we are so close to it. If the sun and Sirius were both the same distance from the earth, Sirius would look *much* brighter.

Big or Small

There is an incredibly large star not too far from us that is called **Betelgeuse** (beet' uhl jooz). It is so big that if it were sitting in the middle of our solar system, it would take up all of the room from the sun to Jupiter! That's many, many times bigger than our sun. In fact, it is so big that astronomers call it s **supergiant** star. You can find Betelgeuse at the top of the constellation **Orion** (oh rye' uhn). Orion is called "the hunter," because if you connect the stars and use your imagination, it kind of looks like a man holding a shield (or maybe a bow).

Stars not only get labeled for how hot and bright they are, but they also get grades on how much total energy comes out of them. In general, bigger stars make more energy than smaller stars, so these grades also tell you how big the star is. The grades are given with Roman numerals, and Betelgeuse gets the highest grade.

I supergiant stars (Betelgeuse is a I.)
II bright giant stars
III giant stars
IV subgiant stars
V main sequence stars (Our sun is a V.)
VI sub-dwarf stars
VII white dwarf stars (The star that orbits Sirius is a VII.)

This is a photo of the constellation Orion. When looking at the sky, the shield and head are very hard to see. However, the body, belt, and feet are easy to find. Without the head and shield, it looks like a big "K" in the sky.

Our sun is a G-5 V star. Can you tell me exactly what that means? Isn't that neat that you can now study about stars and understand the letters and numbers that they are given?

Put into words all that you remember so far. Don't forget about the North Star, Polaris, and the brightest star, Sirius. Also explain what black holes and supernovas are.

Light Years

Did you know that light travels very quickly? Light travels at 186,000 miles per second. Say that out loud. Light travels at a hundred eighty-six thousand miles per second. That means that when you turn on the light, it traveled that fast to light up your room. When you see the light from the sun, the light you are seeing has traveled 93 million miles, going 186,000 miles per second to get here. That means it takes the light from the sun about 8 minutes before it finally gets to your eye. So, when you see the sun each day, you are really seeing what the sun looked like 8 minutes ago! You can never see what the sun looks like right this minute unless you get very close to the sun in a spacecraft. If you were on Pluto, the light you would see from the sun would be about 30 minutes old by the time you saw it.

The stars in the sky are very far away from us. The closest star to us is Proxima (prox' ih muh) Centauri (sen tor' ee), and it is about four **light years** away from the earth. What does "light year" mean? It means the distance that light can travel in a year. Since light travels 186,000 miles per second, it can travel a *long* way (about 5,900,000,000,000 miles) in a year. This means if we were going as fast as light (186,000 miles per second), it would take us four years to reach the closest star! What's really incredible is that most of the stars in the sky are hundreds, thousands, millions, even billions of light years away from us. Even traveling at the speed of light, then, we could not visit most of the stars in our lifetime. Since that is the case, God probably does not intend for us to visit them. Who would want to deal with all their explosions and unpredictable temperatures anyway?

Some people claim that since some of the stars are billions of light years away, the universe (and therefore the earth) must be billions of years old. After all, they think that the light from those stars had to travel from the star to the earth, and since the light is hitting the earth, it must have had billions of years to travel. They do not understand that God is so wise that He could create things to be exactly as He wants them to be. If He could create the first man to be fully grown and fully developed, He could create stars and their light to be fully grown and fully developed as well. He could create an earth with starlight that was already upon the earth, even starlight from billions of light years away. After all, nothing is impossible for God.

There is another reason that the stars can be billions of light years away and the earth could still be quite young. A man named Dr. Humphreys, who is a scientist, wrote a book called *Starlight and Time*. This book tells us that the universe might actually be a lot older than the earth, because the universe around us aged differently than the earth itself.

Dr. Humphreys believes that the whole universe centers around the earth and our solar system. This makes sense, since God created the earth before He made the stars. Dr. Humphreys also believes that during the initial creation discussed in the book of Genesis, time moved faster in the outer regions of the universe than it did on earth. In other words, Dr. Humphreys believes that outer space got older more quickly than the earth and our solar system did. Because of this, even if God did not create the light from stars already shining on the earth, light from the stars would have had time to travel to the earth, even if the earth is very young. After all, if the universe aged more quickly than the earth, the universe could be very old, while the earth is still very young. Although it might seem strange to you that the universe could age differently than the earth, this is actually an idea that most scientists understand. When you get older, you might study something called **relativity**, which explains how this can happen.

Galaxies

Stars are found in groups: big, big groups. These big groups of stars are called **galaxies**. Galaxies are billions of stars all arranged together. Galaxies come in four shapes; **spiral**, **elliptical,** **lenticular** (len tik' you lur) and **irregular**.

This is a spiral galaxy. It is shaped like a spiral, of course.

This is an elliptical galaxy. It is more of a blob.

This is a lenticular galaxy. It kind of looks like a spiral galaxy, but it has no "arms" like the spiral galaxy above.

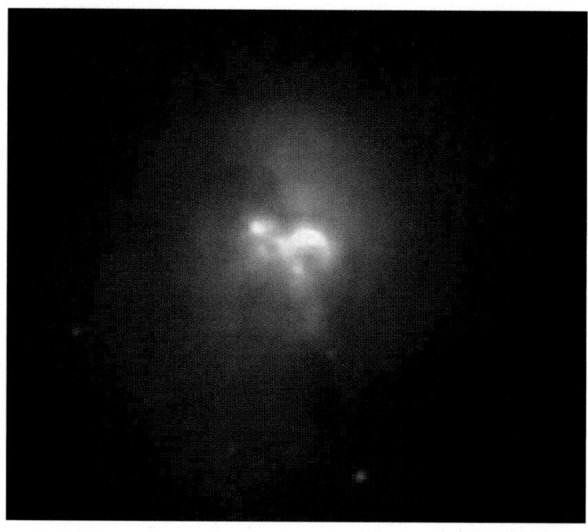

This is an irregular galaxy. If a galaxy is not a spiral, elliptical, or lenticular galaxy, it is an irregular galaxy.

The earth is in a galaxy called the **Milky Way**. Our galaxy is called the "Milky Way" because if you go outside and look up at the sky on a very clear night where there are no city lights nearby, you will see a milky white, almost pink, smear across the sky. That smear is actually one "arm" of the spiral that makes up our galaxy. Our sun is one of billions of other stars in the Milky Way, but we can be certain that God holds it to be the most important star. Unfortunately, many people cannot see the Milky Way anymore because of bright city lights that drown out a lot of the stars.

The Milky Way is a spiral galaxy, and we are out on one of the arms of the spiral. Scientists have determined that the only location in which a star could have a solar system capable of supporting life is on the arm of a spiral galaxy. Our existence is nothing short of a miracle. The earth had to have the perfect size, the perfect rotation, and the perfect environment to support life. The earth needed a medium hot, medium sized, fairly bright sun. We also had to be the perfect distance from the sun. In addition, we now learn that for any life to survive on a planet, it must be in a spiral galaxy, out on an arm. It is amazing that some people think we are all here by accident. The chance of an accident like us occurring anywhere in the whole universe is almost impossible!

Galaxies are so enormous that our mind cannot even imagine how big just one galaxy is. There are billions of stars in a galaxy, and there are also billions of galaxies in the universe. Some are so far away that astronomers can barely see them. Remember that things that are far away seem smaller. Some of the galaxies that astronomers have noticed seem so small that each entire galaxy looks like a speck of glitter on the telescope. When they study these specks so far away, it is hard to understand what they are seeing. It is very hard to know exactly what is going on, even in our own galaxy, not to mention a galaxy that is billions of light years away!

Explain what you know about galaxies in your own words.

Constellations

The heavens are telling of the glory of God; And their expanse is declaring the work of His hands. Day to day pours forth speech, And night to night reveals knowledge. ***-Psalm 19: 1-2***

When you look up at the sky in November, you will not see the same stars that you saw in May. Every month the stars we can see change. That is because the earth is revolving around the sun and is therefore facing different sections of the universe each month. There are stars all over the universe, and as we orbit the sun, the night sky shows us different views of the universe.

As I mentioned in Lesson 1, many of the stars form patterns in the sky, and we call those patterns constellations. On the next page, I will show you show you pictures of some of the constellations.

This is the constellation Cassiopeia (ka' see uh pee' uh). It looks like a deformed "W" in the sky.

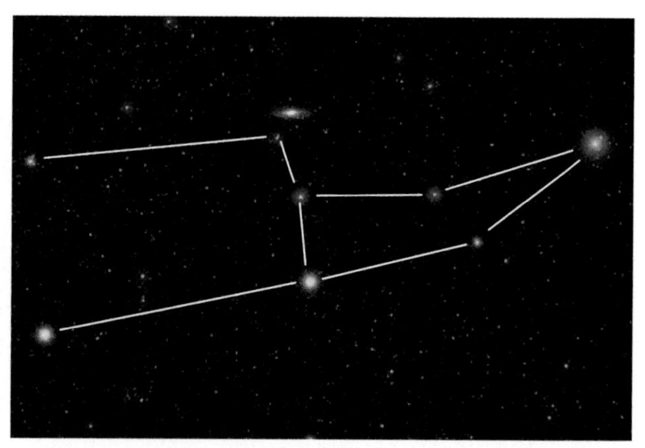

This is the constellation Andromeda (an drah' muh duh). It looks a bit like an "A" on its side.

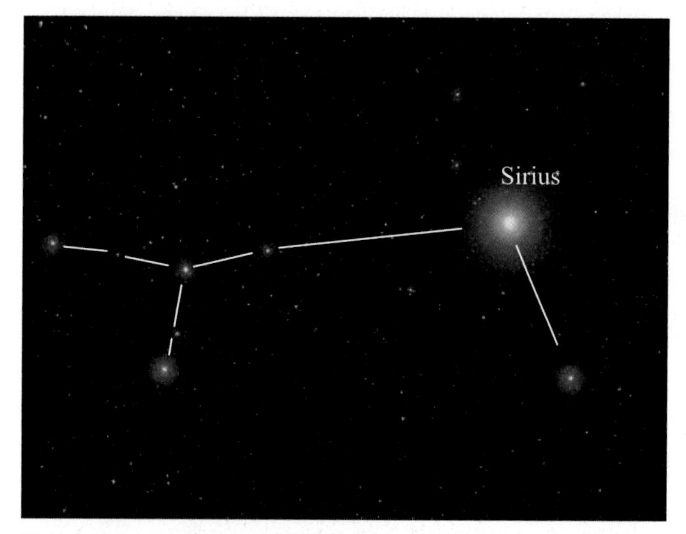

This is the constellation Canis Major. It is supposed to look like a dog, because "canis major" means "great dog." It is easy to find because the brightest star in the sky is in it.

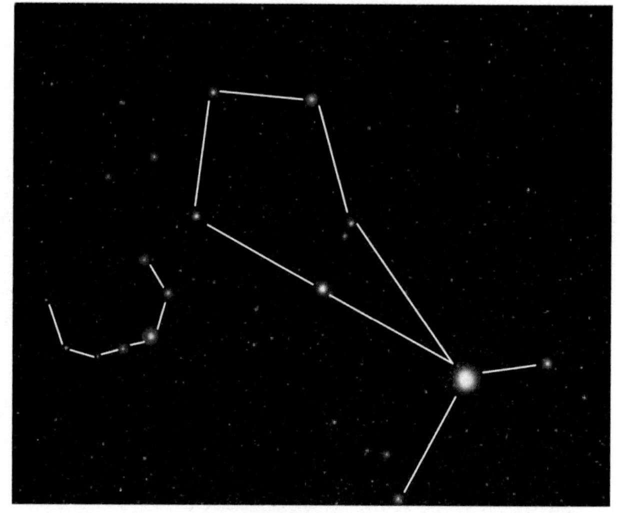

There are two constellations here: Bootes (boh oh' teez) and Corona (kuh roh' nuh) Borealis (bor' ee al' us). Corona Borealis looks like a backwards "C" in the sky.

Now remember, you have seen a picture of the constellation Orion on page 148, and on page 143, you saw a picture of the Big Dipper and the Little Dipper. The Big Dipper is actually a part of a constellation called **Ursa Major**, which means "great bear," and the Little Dipper is a part of the constellation **Ursa Minor**, which means "little bear." The Big Dipper and Little Dipper are much easier to see than the entire constellations, however, so it is best to start out looking for the dippers rather than the bears.

Please realize that the constellation pictures you see in this book have been enhanced to make the patterns easier to see. When you actually look at the night sky, the constellations will not be as easy to see. However, with a little patience and practice, you will be able to find them. You can also get some help by going to the course website that I told you about in the introduction. There will be links to star maps there that will help you know where to look for each constellation.

Gospel in the Stars?

Although some people used to tell stories about their false gods using the constellations, the one true God is the one who created the constellations. God mentions them several times in the Bible as something He created. Some Christians believe that God originally intended the constellations to symbolically tell the story of Jesus, our Savior! God says in the Bible, "Can you bind the chains of the Pleiades, Or loose the cords of Orion? Can you lead forth a constellation in its season, And guide the Bear with her satellites?" (Job 38:31-32). The Pleiades, Orion, the bear, and the bear's satellites are constellations. The bear and its satellites are the constellations Ursa Major and Ursa Minor. Remember, the Big Dipper is part of Ursa Major, and the Little Dipper is part of Ursa Minor. Do you remember which star is the last star in the handle of the Little Dipper? It's Polaris, the North Star.

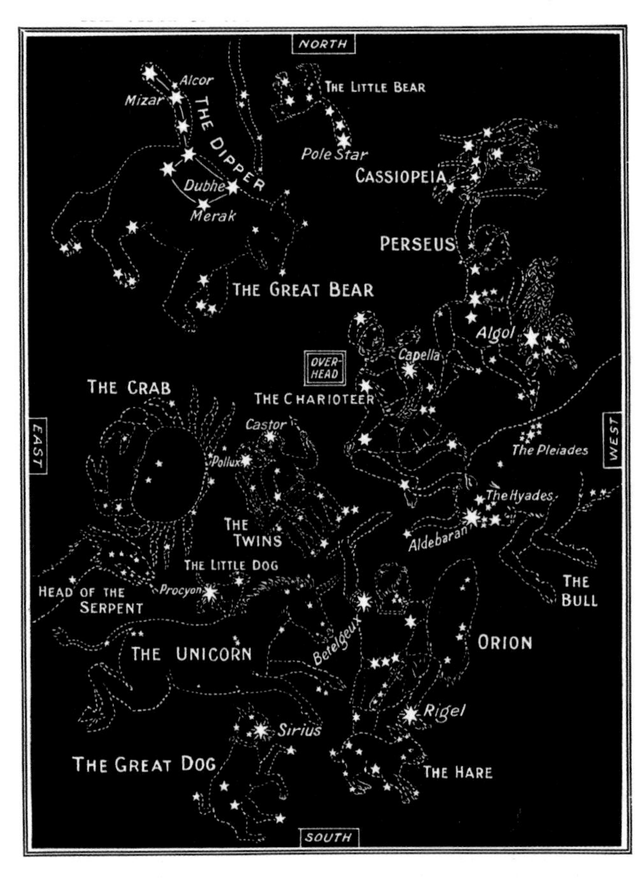

In the above Bible verses, God is telling Job about the constellations that He created and intended for us to know. Historians know that the first people to tell stories about the stars by constellations were God's people: the ancient Hebrews. Did God show the story of His salvation through Christ in the stars? The constellations have symbols that God uses in the Bible; a lion (Jesus is called the Lion of the tribe of Judah,) a virgin (Jesus was born to a virgin,) a serpent (Satan is portrayed in the Bible as a serpent), a king, and many others. Many believe Orion points to Christ, the Conqueror.

Thankfully, we do not need the constellations to tell us of the coming Savior. He has already come, and we now have His written Word, the Bible. Are you telling others about Jesus and showing them how to understand what Jesus says in the Bible? That is what God asks all of His children to do.

Why would God tell stories with the stars in the sky? Well, people didn't always have the Bible the way we do now. They didn't have words written on paper to remind them of God's purposes. We don't need the stories in the stars anymore because we have God's written word! Praise God! Don't forget to read your Bible; it's the best book in your house and the only one that is a living book. Remember, it is the very Word of God.

Corruption of Truth

Sadly, Satan is always trying to turn mankind from the one true God and trick us into worshipping idols. The people who didn't love God made up their own stories in the stars to go with their own religion. They made the stories in the stars to be about idols. Constellations kept the same pictures that God's people called them, but Satan deceived mankind into removing the true hero, Jesus Christ, from the stories and inserting lies with false gods.

Satan has also created a religion where people put their faith in the stars, believing that where the constellations are in the sky has power over their lives and their future. That's silly, I know, but it is really true. This is idol worship. It is putting your faith in the creation and not the Creator. This religion is called **astrology** (uh strahl' uh jee). It sounds a little like astronomy, but it's very, very different. Astrology is an old religion, and is still practiced today, even in America! In this religion, people check their 'horoscopes' to see what is supposed to happen to them each day. Some people are serious about it, and some people are not. But all idolatry is serious to God. They need to learn about Jesus, don't they? We need to tell them the truth about the stars and God's purpose for them.

God's plan for the stars is for His glory and not for man's. God's desire is for us to glorify Him and not the creation He made. If you ever meet anyone who follows astrology, God would have you share the truth about Jesus with him or her.

What are constellations? Can you explain them in your own words?

Constellations and Astronomy

Why are constellations still given importance in astronomy today? They help people, both astronomers and stargazers, know where they are looking. In other words, it gives them a map of the sky and helps them to find what they are looking for. If you wanted to find Betelguese, for example, you would first look for the constellation Orion, and then it would be easy to locate Betelgeuse.

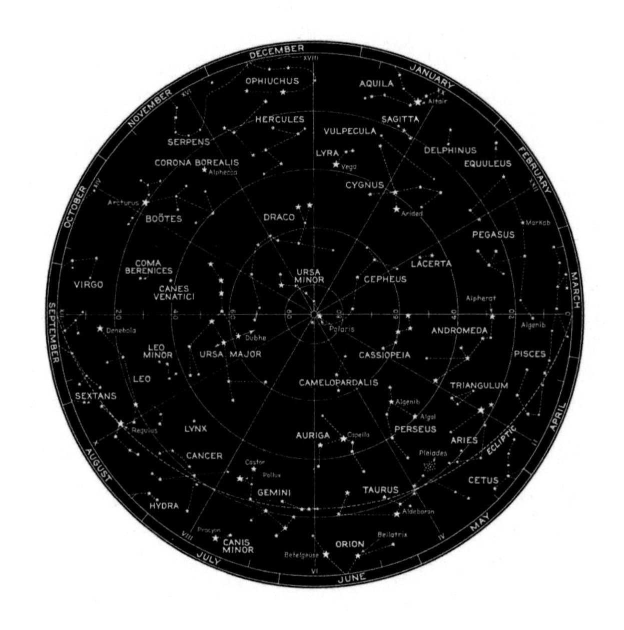

Before people had a compass, they knew which direction to travel by land or by sea if they looked up at the constellations. They knew which way was north, south, east, or west just by the stars.

You can purchase star maps or planispheres (plan' uh sfears) to know exactly where the constellations are. They will help you to find the constellations in the sky. If you do purchase a planisphere, you must get one that is for the region where you live. Typically, planispheres are ordered for a particular **latitude**, which marks how many degrees you are from the equator. If you live in the Southern United States, you will want a planisphere for a latitude of 30° north. That means you are 30 degrees north of the equator. If you live in another part of the United States, find your state on the map to the left and determine which latitude you are closest to. If you are not in the United States, you can get your latitude from an atlas or online at www.worldatlas.com.

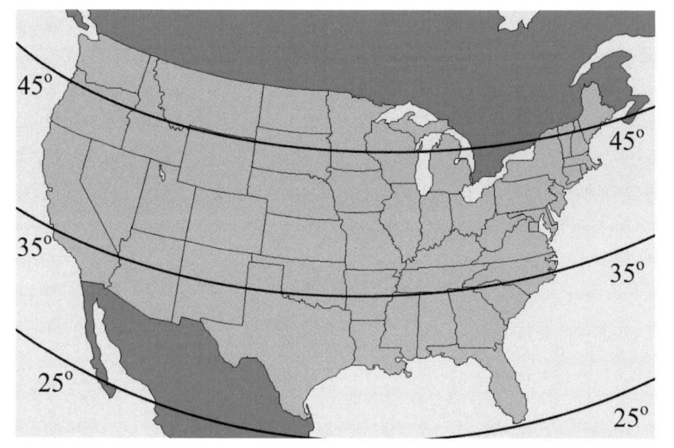

What Do You Remember?

Why do you see different stars during different times of the year? Which group of stars is always present in the night sky of the Northern Hemisphere? What is the name of the North Star? What is special about the star named Sirius? What is a black hole? What is a supernova? Describe the three star categories. What is a galaxy? In which galaxy is the earth? What is the shape of our galaxy? What is a constellation? How are constellations used today? What is the difference between astronomy and astrology?

This incredible photograph, taken by the Hubble Space Telescope, shows two galaxies that are actually colliding! Scientists think that as time goes on, these galaxies will actually merge and form one new galaxy. Can you tell what kind of galaxies these are (irregular, elliptical, spiral, or lenticular)?

Assignment

Illustrate a page in your notebook for stars and galaxies. Write down all that you remember about stars, supernovas, black holes, and star categories. Also include what you remember about galaxies, especially the Milky Way.

Make a mnemonic phrase for the star categories that are based on temperature. Do you remember that they are placed in order from hottest to coolest? The letters are O, B, A, F, G, K, and M.

O	**B**	**A**	**F**	**G**	**K**	**M**

Place your mnemonic phrase in your notebook.

Activity

On a nice, clear night, go outside and try to find some of the constellations in the sky. The constellations that you will be able to see will depend on where you live and what time of year it is. To find out what constellations are visible for you, go to the course website I told you about in the introduction. There, you will find links to places where you can get star maps for your area and time of year. Those star maps will tell you roughly where to look to find certain constellations.

The best thing to do when you are looking for the constellations is to find someplace very dark. If you live in the city, you might try driving out to a country field or a park to get away from the city lights. Carry a tiny flashlight with you. If you have red plastic, like colored plastic wrap, cover the flashlight with it so that the light will be red. That way, when you use your light, it will not affect your eyes as much.

Before you go out, study your star map for a moment to try to get an idea where you will look for what constellations. Once you get outside, lay on a blanket or reclining chair and simply look at the sky for a while. This will get your eyes used to the dark. Then, look for the constellations. If you need to refer to your star map, do so only with the tiny flashlight. The darker you can keep it, the easier it will be to see the constellations, because your eyes will be more used to the dark.

Project
Make an Astrometer

Do you remember that our sun is a G-4 star? Remember that "4" means it is a pretty bright star. You will make an astrometer (as troh' mee tur) that will measure the brightness of the stars according to the numbers given stars by astronomers. Our astrometer will measure grades 1-4. Please note that because we are on earth, our astrometer will measure the *apparent* brightness of the stars (how bright they are from earth), which is a bit different than the *actual* brightness of the stars. If you ever study astronomy in college, you will learn how to take the brightness you see here on earth and turn it into the actual brightness of a star.

You will need:

- Cardboard
- Clear plastic wrap
- Tape
- Scissors
- A marker

1. Cut four rectangular slots in the cardboard.
2. Tape a long piece of plastic wrap over all four rectangles.
3. Tape another piece of plastic wrap over the three top rectangles.
4. Tape another piece of plastic wrap over the two top rectangles.
5. Tape another piece of plastic wrap over only the top rectangle.
   Now you have four rectangles covered in plastic wrap. One has only one layer of plastic wrap over it, the next has two layers, the next has three layers, and the last has four layers.
6. Write a "1" next to the rectangle with four layers of wrap. Write a "2" next to the one with three layers, a "3" next to the one with two layers, and a "4" next to the one with only one layer of wrap. You have just made an astrometer. You can use it tonight to measure the brightness of stars.
7. Take your astrometer outside on a clear, dark night, and begin by looking through square number four. You will see many stars through it.
8. Look through square number three. You will see fewer stars this time.
9. Look through number two, and move to number one. The rectangle with the most plastic wrap only shows the brightest stars in the sky.

If you want to determine the brightness of a single star, look at it through each rectangle, starting at rectangle 4 and progressing through the rectangles in decreasing order. The number of the last rectangle through which you can still see the star gives you the brightness of the star.

Project
Create a Constellation Planetarium

You will need:
- A shoe box
- A marker
- A bamboo skewer or ice pick
- A flashlight
- Scissors
- A dark room

1. Cut a hole in one side of the shoebox. The hole should be just big enough to fit your flashlight through it.
2. On the side of the box opposite the hole, make dots in the shape of the Big Dipper. If you have room, you could make constellations as well. Refer back to the pictures in this lesson to see what the constellations look like.
3. Use the bamboo skewer to poke holes in the box where your marks are.
4. Find a room that can be darkened well. Close all of the curtains so that the only light comes from the light bulb in the room.
5. Place the flashlight in the hole that you cut for it, making sure that its light will shine into the box. You may need to place a book under the flashlight to support it.
6. Bring the box pretty close to the wall, and point it so that light from the flashlight will shine through the constellation holes and onto the wall.
7. Turn on the flashlight.
8. Turn off the room lights. The patterns you made in the box will appear on the wall.

Lesson 14
Space Travel

Let's Go to Space

Since the beginning of time, man has been fascinated with God's starry hosts, the universe. When man is interested in something, he usually wants to see it up close. For thousands of years, men have dreamed of visiting the planets. So far, no one has ever been to another planet. That's still a dream. Nevertheless, man has accomplished a lot in his explorations of the universe, and I expect that a lot more will be accomplished in the next generation.

People who study ancient history learned that back in the 1200s, the Chinese had figured out how to build rockets, but they did not share how to do this with others. Many people tried to make rockets that would blast them up into space. People used dynamite or gunpowder to try to launch rockets into space, but they had no success. In the late 1800s, a Russian astronomer, Konstantin Tsiolkovsky (sil' oh kof' ski), believed that man would need to mix liquid hydrogen and liquid oxygen to make a fuel strong enough to allow a rocket to get away from the earth's gravity. No one tried it until many years later, but it did work. Today, we use a very similar mixture to blast rockets into space! Because of this, Tsiolkovsky is called the "father of astronautics (as truh not' iks)." The word **astronautics** means "the science of space flight."

Later on, in 1919, a man named Robert Hutchinson Goddard studied how a person might get to the moon. He told people that a rocket should be used. Everyone laughed at him, made fun of him, and called him names, like "Moon Man," because they thought that was the strangest and most ridiculous idea they had ever heard. But that didn't stop him from working on his rockets. He made them so that they were fast! They could travel with a speed of about 700 miles per hour, and they could go about a mile and a half up into the air before falling back down to the ground. Remember that our atmosphere is about 800 miles high, so Goddard had to keep working to make rockets that would get out of the earth's atmosphere.

This is a photograph of Robert Goddard.

About the same time, Germany was in a war and figured out how to make airplanes fly 100 miles per hour and 3 miles up into the air. Man was getting higher into the atmosphere. It wouldn't be long before they sent a rocket into space. In fact, many years later, the Russians did just that! Can you find Russia on a map? It's a very large country. Many people back then were frightened of the Russians because their country was not free, like the U.S. and other countries were. People could not live anywhere they liked or do what they chose to do in Russia. They were told what they could and could not believe. At that time, the Russian government taught the Russian people to not believe in God. They controlled their lives in a way that frightened free nations.

Sputnik Sensation

The Russians used a lot of ideas that the Germans developed during World War II to make a rocket that would launch into space. The rocket's job was to put the first artificial satellite, called **Sputnik**, into orbit around the earth. This was the first rocket to go to space, and Sputnik became the first artificial satellite. The U.S. and other countries were worried about this Russian rocket, because many countries were not friends with Russia. Many nations were scared that Russia might become very powerful indeed, taking over the whole world with such amazing, advanced technology.

This is a photograph of Sputnik 2 about to be launched into space.

As the U.S. was scurrying about to build a rocket of its own, Russia put another rocket into space. Sputnik 2 was even more amazing because it carried a little dog into space with it. The dog was named Laika (like' uh), which means "barker" in Russian. The cabin Laika lived in was pressurized. That means it had an artificial environment with oxygen and heat. Before Laika went to space, many scientists believed that living creatures could not survive without gravity and the protection of our atmosphere. Laika showed the world that living things could survive in space. The Russians were really doing amazing things!

Sputnik created a big 'problem' for the United States and other free nations. They were afraid of Russia getting too much power and taking over the world. In response to all the Russians were doing, the U.S. began to put together a big space program to try to get ahead of the Russians. They called this the "**Space Race**" between the U.S. and Russia. This is what started all that we know today about space and space travel. Just a few months after Sputnik, the U.S. launched Juno 1, a rocket carrying a satellite called Explorer 1. Phew! Finally, a nation besides Russia had a satellite too. A few months later, NASA (National Aeronautics and Space Administration) was born.

The 1960s

The 1960s are known for great discoveries in astronautics, the science of space flight. The Russians put an actual human being in space in April of 1961! The Russians called this man a "cosmonaut." A month later, the U.S. sent Alan Shepard into space, and Americans called him an astronaut, instead of a cosmonaut.

When Shepard got back to earth, the president of the United States, John F. Kennedy, decided that the U.S. should try to put a man on the moon. No one laughed at him the way they did Goddard forty years before. Eight years after President Kennedy publicly stated his desire to put a man on the moon, an Apollo spacecraft sent Neil Armstrong to the moon. The United States had finally done something first! They were very excited. Do you know what Neil said when he stepped on the moon? He said, "That's one small step for man, one giant leap for mankind."

This is a photograph of Neil Armstrong on the moon.

The United States beat the Russians to the moon, but the Russians beat the Americans again a few years later by building the first space station. The Russian space station was called Salyut. What is a space station? It's a giant satellite that people can live in. The living area on the first space station was smaller than many people's closets. Many of these small space stations were built and sent into space by the Russians and the Americans. Some were a success, and some were unsuccessful and fell apart while on their way into space.

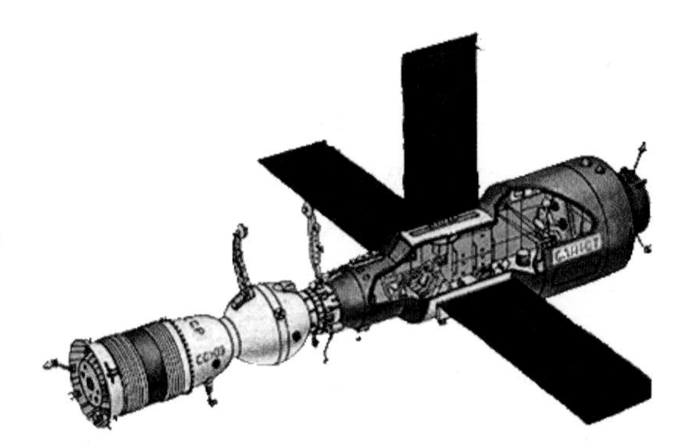

This is a drawing of the Salyut space station.

Many years later, the Russian government fell apart, and the U.S. was no longer in a race with them. Americans and Russians are now friends and help one another to learn about space and astronautics.

Space stations are a great way for scientists to study space. Space stations orbit around the earth so quickly that they can look at much of the universe in a few hours. In fact, night comes about every hour on a space station. That means space station astronauts get to see the side of the earth having night and the side of the earth having day in about two hours!

The greatest and most important use for space stations is that science experiments can be done in space! I'm sure you're wondering why someone would want to do experiments up in a little space station rather than right here on earth. The reason is that when experiments are done without the influence of gravity, the results are different. Gravity affects everything: how plants grow, how fire burns, how germs spread, how chemicals mix together. The experiments done on the space station help us better understand how things work in the absence of gravity. This helps scientists make better medicines, equipment, products, and even toys!

I bet you didn't know that NASA makes toys! NASA science has led to lots of toys. Because NASA understands how to make things fly well, a toy company hired them to build a toy glider that would go a long way before hitting the ground. NASA did a good job. NASA also developed the controllers that many children use to play video games on their TVs. These controllers are just like the ones used to fly spaceships to the moon!

Space stations are pressurized with oxygen, heat, air conditioning, electricity, and many of the comforts of earth. Inside the space station, the astronaut or cosmonaut doesn't have to wear a spacesuit. He is protected from the harsh environment outside. It's just like a home away from home.

The International Space Station

The best space station ever built was assembled in 1998. This space station is called The International Space Station. "International" means "more than one nation." NASA used the word "international" because many nations helped NASA build the International Space Station, including Russia.

This is a photograph of the partially-completed International Space Station orbiting the earth.

The International Space Station is like a home and science laboratory all in one. The people living on this space station are trained astronauts. They were hired because they are scientists who do experiments with plants, chemicals, and many other things. Do you like to do experiments? Perhaps one day you will do your experiments in space.

NASA is always looking for expert scientists to do experiments in space. They need people who know a lot about life science, earth science, space science, gravity science, engineering, and space product development (making clever things). Almost any area of science can be useful to NASA.

The scientists living in the space station have to get used to all kinds of changes from life on earth. It takes them only 90 minutes to orbit the earth, so the sun rises and sets every 45 minutes. With the sun rising and setting like that, deciding when it's bedtime is a little more difficult. God made us to be on a 24-hour schedule. We operate best on the schedule God created for us, so even the astronauts try to keep a 24-hour schedule every day.

What is it like to live with no gravity on the space station? Scientists can simulate, or imitate, almost everything you will feel in space. However, on earth it is very difficult to simulate not being affected by gravity. You can do it for short times by being in an airplane that is falling. While the airplane is falling, you feel no gravity. In fact, that's why the astronauts don't feel any gravity in the space station. It's not that the earth's gravity doesn't exist there. It certainly does. That's what holds the space station in orbit. However, the astronauts don't *feel* the gravity, because for them, it is like the space station is always falling.

I know this is tough to understand, but let me try to explain it. Suppose you were on an elevator, and the cable that holds the elevator suddenly breaks. Oh no! The elevator would start falling down the shaft, and you would be falling right along with it. Do you think that your feet would touch the floor of the elevator during the fall? They would not. You see, both you and the elevator are falling. Since you are falling at the same speed, you can never reach the floor of the elevator. As a result, you would float in the elevator, as if there were no gravity. That's what happens on the space station. The space station is constantly falling as it travels around the earth. The astronauts are falling with it. Because of that, the astronauts float, experiencing no gravity.

This astronaut is floating because both she and the spacecraft she is in are orbiting the earth. This means both she and the craft are falling, so she experiences no gravity, even though gravity holds the craft in orbit.

No gravity is called zero gravity, or "0-gravity." It is impossible to really prepare someone for experiencing zero gravity over a long time. Astronauts can experience zero gravity for a short time by getting in a plane and allowing the plane to fall to earth for a while. However, they can only do that for a short time, because they have to get the plane flying again before it hits the ground. To try and prepare the astronauts for longer exposure to zero gravity, NASA trainers put the astronauts in tanks of water. Since the water makes you float, it is a bit like experiencing no gravity. However, it's not exactly the same because you can control your movement by pushing against the water. In zero gravity, there is nothing to push against to keep you steady or to help you move where you want to go. You simply float, like a balloon thrown into the air.

This astronaut is experiencing 30 seconds of zero gravity using a NASA craft that goes up into the air and then falls towards the ground. At the last moment, it pulls up to avoid crashing. This craft is often called the "vomit comet," because falling like that can make people sick at first.

In space, an astronaut can fly and tumble around like Peter Pan. It's hard at first to get used to, but once the astronaut settles in, life in space just seems normal. An astronaut gets used to not having any weight. The moment an astronaut returns to earth, however, the astronaut's body suddenly feels like it weighs as much as twelve trucks. That's because she is not used to feeling gravity pull on her. To suddenly have the force of gravity pulling on you so hard that you would be pulled right down to the ground from miles up in the air must be a strange feeling.

Gravity keeps everything in place and working as God designed it. Did you know that your blood flow was designed to use the force of gravity? On earth, gravity keeps the blood flowing down to your legs. We need a lot of blood in our legs because they do so much work for us. In space, our legs aren't as important anymore. The astronaut hardly uses them at all. He must get lots of exercise to keep them from becoming too weak to support him when he comes back home. Without gravity, the blood gathers into the chest and upper body, instead of flowing down into the legs and feet. That's why astronauts in space can get large upper bodies and skinny legs. This is called chicken-leg syndrome. Lots of exercise helps to avoid this, so astronauts exercise each day they are in space.

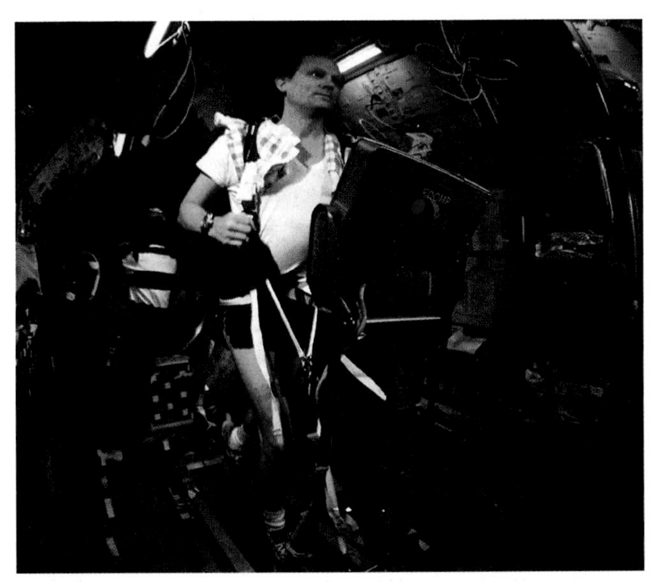

This is an astronaut exercising on the Russian space station Mir. Notice the straps that are holding him. They keep him from floating off the treadmill.

Did you know that water and other liquids don't flow downward in space? Just imagine if you spilled your milk while having lunch on the space station. Your milk would float about everywhere! Liquids ruin electrical equipment. Astronauts have to be very careful with liquids in space. They get drinks by sucking water out of a water bag that looks like a juice bag. They can also get water from a tube attached to the wall. Astronauts can't set their food on a plate. It will simply float off. Food is kept in pouches and eaten from the pouches. It is usually dried, with no crumbs. Imagine spilling crumbs in space! They would float all over the place, in every direction. It would be difficult to clean up, and it might get in your equipment and ruin it! The equipment on the space station is very delicate. Astronauts must take special care of the space station. They depend on everything to work perfectly so that they can stay alive and well. Crumbly food is not allowed.

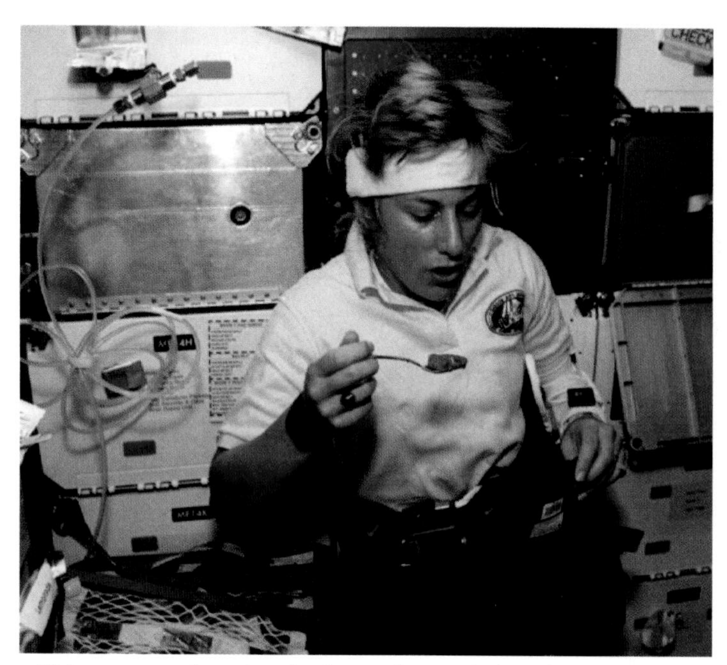

This astronaut is eating food out of a pouch that she is holding.

Fresh fruits and vegetables are also sent up to the astronauts in the space station, but it is very expensive to send shipments to space. It's not like sending a package to Grandma: the delivery man must be a trained astronaut, and millions of dollars are spent launching the delivery out of our atmosphere to the space station. Getting supplies of food, water or anything else is also very tricky because the space station is moving quickly in its orbit around the earth. The spacecraft making the delivery must be guided very carefully so that it can dock with the space station. Shipments aren't sent up very often, so the space station must have a lot of supplies on hand.

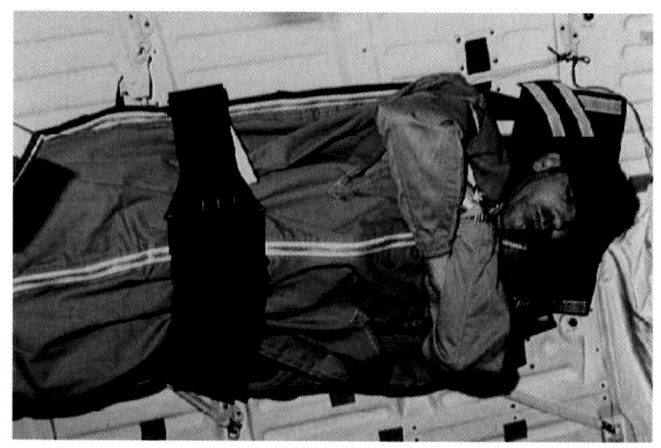

This astronaut is sleeping. It doesn't look like a very comfortable "bed," does it?

Astronauts on the space station sleep in sleeping bags that are attached to the wall. They zip themselves in and strap themselves down so they will not float away in the middle of a dream.

Using the restroom is also a lot different in space. Our toilets use gravity to work. Flushing the toilet only works because gravity pulls the water down. Since the contents of the space station do not feel gravity, there is no "down." In space, the toilet is attached to you and gently sucks away all your waste. That may sound awful to us, but to astronauts, it's just a normal part of life.

If water droplets somehow get into the air, these water droplets must be caught or they could ruin the equipment. Water floats into the air, like big water-filled balloons that will break on whatever they hit. Special vacuum cleaners suck the water out of the air.

Believe it or not, astronauts use a squirt gun to take a shower! Taking a shower requires several people. One person is in the shower with a squirt gun while the others stand outside the shower. The person taking the shower squirts water on himself while his friends stand outside with the water vacuum and suction up the water that floats out above the top of the shower. Showers are so difficult that astronauts usually just take "sponge baths" with a wet washcloth. That is much easier.

Not knowing which way is up can bother the astronauts a bit. Because of this, space station builders, or engineers, have placed drawers, closets, equipment, and signs in such a way that the astronaut knows which way is "supposed" to be up. Astronauts need to know this, because they are often upside down, sideways, and doing tumbles in the air. An astronaut could just as easily walk along the ceiling as on the floor.

Building the International Space Station

Much of the International Space Station was put together up in space. The pieces, called **modules**, were built on earth and sent up to be put together in space. Rockets and space shuttles, loaded with parts and pieces of the space station, were launched into space, so that the materials could be delivered to astronauts who are trained to build the space station.

This is a photo of the first space station module (called Zarya) being sent into space. Zarya is the control module.

Launching the space station modules up to the space station was very expensive. It is a difficult task to get them out of our atmosphere. Once in orbit, however, the giant pieces of the space station become weightless! You can lift a module piece, weighing as much as a building on earth, like it weighs nothing at all.

Leaving the space station to work on the outside is called a **space walk** or an **EVA**. "EVA" stands for "extravehicular activity." Space walks are very, very dangerous. When an astronaut leaves the inside of the station to work outside, the temperatures are extreme. Without an atmosphere to keep the temperature stable, it is burning hot when the sun shines upon the space station, and it's terribly cold when the sun is not shining. It's also quite dangerous with all the solar winds from the sun beaming harmful radiation right at the astronaut. Do you remember about the shield around the earth that keeps the solar winds directed toward the North and South Pole? That shield is called the magnetosphere. If an unexpected solar wind came suddenly, it could harm the astronaut very badly. Also, there is no protection from all the dust and rocks hitting our atmosphere nonstop day and night. Those little meteoroids continually rain on space-walking astronauts, like a constant sandstorm. Some of the meteoroids are as big as a golf ball! Even a pebble-sized meteoroid flying at you could cause a severe injury, but the spacesuits worn by astronauts protect them from as much danger as possible. Despite all the danger, space walkers enjoy the excitement of being out of the ship.

This is a photo of an astronaut waving at the camera while he is on a space walk outside of the International Space Station.

You need to have a lot of training to be a space station construction worker. Building a laboratory and home in space must be done perfectly, or it could be dangerous for those living in the space station. This is hard work in a big bulky spacesuit, with no gravity to keep you stable. If the astronaut wants to turn, he must move his whole body. Otherwise, he might lose his balance and float away from the space station. That's why there is usually a line that is attached to the astronaut and the space station. It keeps the astronaut from floating away and into space. The astronaut building the space station has a dangerous and important job.

Becoming a NASA Astronaut

It would be a sad thing if some children didn't want to become scientists when they grew up. It would also be sad if they started out wanting to be scientists but then gave up when the classes became difficult. Being an astronaut is great fun. You might think you have to study space and astronomy in school to become an astronaut. Actually, NASA hires lots of people to become astronauts. Medical doctors, scientists who work in all sorts of different fields, and all kinds of engineers are employed by NASA. There are so many areas of science that are needed in space studies that many fields of science are good preparation for becoming an astronaut.

If you want to be an astronaut, it is important to take lots of math and science classes in school. Don't rush this part. Take your time and learn about the earth, its rocks, and its plants. Study living things, animals, and how bodies work. Study energy, motion, and electricity. All of these things are important fields of science, and NASA needs people trained in all of them. After you have studied many areas of science, decide which is your favorite. You will become an expert sooner than you realize! If your dream is to become an astronaut, you can do it with a lot of hard work and prayer. You are a special child of God and have the power of Jesus within you to do things that are difficult. As it says in the Bible, you can do all things through Jesus who gives you strength (Philipians 4:13).

This is a photo of an astronaut clowning around. He has just completed a space walk in which he successfully retrieved two satellites that were not working. They were loaded into the space shuttle to be brought back to earth. I guess he was trying to make a little extra money, but there aren't many potential buyers up there.

Don't forget your first and most important goal is to bring honor to God and glorify Him, because all things in heaven and on earth were made through Jesus and for Jesus. Whatever you do, then, do it as if you are working for Jesus and not for people. If you do that, you will be fulfilling God's purpose for your life.

Seeing the International Space Station

Did you know that you can actually see the International Space Station in the night sky? It is close enough to earth that it appears like a big star. However, this "star" moves very quickly across the sky, because it orbits the earth once every 90 minutes. If you go to the course website I told you about in the introduction, you will find a link that takes you to a NASA program. This program will tell you when the International Space Station will be visible in your area and where to look for it.

What Do You Remember?

What was the name of the first artificial satellite? What was the race called between Russia and the United States? Why was the U.S. worried about Russia's space program? What did the U.S. do first? What did Neil Armstrong say when he stepped on the moon? What is a space station? What is the name of the best space station? What is the job of people who live on this space station? What is life like on the space station? How do you become an astronaut?

Assignment

Illustrate a page in your notebook for space travel. You might want to draw the International Space Station, a rocket firing into space, or an astronaut on a space walk. Write down all you remember about space travel and space living. Be sure to write down how you can become a NASA astronaut.

Activity

Let's Visit the Planets!

Older Students: Do the following math exercise on your own to calculate how long it would take you to pay a visit to every planet in our solar system.

Younger Students: Parents, this is a fun activity to do with your child. Get out a calculator!

Instructions

To travel to the planets takes a long time, even in a very fast spacecraft. We are going to do some math to figure out how old you would be if you started visiting the different planets in our solar system. Let's suppose your spacecraft can travel at 10,000 miles per hour. That would be a very fast spacecraft, indeed. Let's also suppose you visit every planet in the solar system in that spacecraft. Follow the directions on the next page to find your age when you reach each planet. You will be surprised at how long you will be traveling!

Imagine you are leaving on the day of your next birthday. How old will you be then? That will be your starting point. We will see how old you will be when you reach each planet. Then, we'll figure out how old you will be when you get home again.

Your flight will begin on earth. You will start by traveling to Mercury. Then, you will turn around and visit all of the planets in order. To make the calculation possible, I will assume that each planet is perfectly placed in its orbit so that you can always travel in the same straight line. In reality, you might have to travel a lot longer than what I calculate here, because one planet might be on one side of the sun, and the next planet might be on the other side of the sun. This would take a *lot* more time.

Starting from the earth and traveling at a speed of 10,000 miles per hour, this is how long it will take to reach each planet:

The day you leave the earth is your next birthday.	How old are you the day you leave?
You arrive on Mercury in 8 months.	How old are you when you get to Mercury? (your age + 8 months)
You arrive on Venus 4 months later.	How old are you when you reach Venus? (your last age + 4 months)
You arrive back on earth 4 months later and say hello to your family before you immediately set off for Mars.	How old are you when you reach earth? (your last age + 4 months)
You arrive on Mars in 7 months.	How old are you when you reach Mars?
You arrive on Jupiter 3 years and 11 months later.	How old are you when you reach Jupiter?
You arrive on Saturn 4 years and 7 months later.	How old are you when you reach Saturn?
You arrive on Uranus 10 years and 2 months later.	How old are you when you reach Uranus?
You arrive on Neptune 11 years and 7 months later.	How old are you when you reach Neptune?
You arrive at Pluto 10 years later.	How old are you when you get to Pluto?
It will take 40 years and 10 months to make it back home.	How old will you be when you get back?

Project
Build a Model Space Station

This project requires a lot of creativity, tape, glue, and stuff you can find around your house. You are going to build a model of the International Space Station.

You can use wires, paper towel rolls, empty soda bottles, craft sticks, straws, lids, and anything else you think might make a nice space station model. Try to make your model look as much like the International Space Station as possible. When you are finished, send Apologia Educational Ministries, Inc. a picture, and we will publish it on the course website! Be sure to include your name and age when you send us your picture.

This is an artist's idea of what the International Space Station looks like when it is in orbit around the earth.

Answers to the Narrative Questions

Your child should not be expected to know the answer to every question. These questions are designed to jog the child's memory and help him put the concepts into his own words. ***The questions are highlighted in bold and italic type***. The answers are in plain type.

Lesson 1

Why did God create the stars and planets? As a calendar, to help the birds know when to fly south, to keep the earth stable, for His own glory. ***What are the names of the planets?*** Mercury, Venus, earth, Mars, Jupiter, Saturn, Uranus, Neptune, Pluto. ***What is the name of America's Space Program?*** National Aeronautics and Space Administration (NASA). ***What does NASA do?*** Builds equipment for studying space, trains astronauts, and sends spacecraft into space. ***Do you remember the name of the astronomer who first said that the earth revolves around the sun?*** Copernicus. ***What about the name of the astronomer who learned how to study space with a telescope?*** Galileo.

Lesson 2

Do you remember how many earths would fit inside the sun? One million. ***How many miles away is the sun?*** Roughly 93 million miles. ***What is the solar system?*** The sun and all the objects orbiting the sun. ***Explain what sunspots are.*** Places on the sun that are cooler than the rest of the sun. ***Do sunspots help us at all?*** Yes, they make the temperatures on earth cooler or warmer, depending on the number of sunspots. ***Does the sun have a satellite?*** Yes, everything orbiting the sun is a satellite. ***Can you explain the difference between revolving and rotating?*** Revolve means to go all the way around, orbiting an object. Rotate means to spin, which turns night into day. ***How does the sun tell us that there were not living things on the earth billions of years ago?*** A billion years ago, the sun would have been too cool to support life. ***Do you remember why you see color?*** Different objects reflect different colors of light that bounce into your eyes. ***Which color has short waves?*** Blue. ***Can you explain what a solar eclipse is?*** When the moon gets in between the sun and the earth, the moon can block the sun's light, making it dark even during the day.

Lesson 3

Can you explain why the sun always appears white on Mercury? There is no atmosphere on Mercury, so there is nothing for the light to bounce off. This means you see all colors of light coming from the sun, which makes it look white. ***Can you explain why Mercury is so cold at night even though it is right next to the sun?*** Since it has no atmosphere, there is nothing to hold the sun's heat in. When the sun is no longer shining, then, it gets cold. ***How do you think it affects Mercury to be closer to the sun?*** It is warmer then. ***How do you think it affects Mercury to be farther away?*** It is cooler then. ***How long is a day on Mercury?*** 59 earth days. ***How long is a year on Mercury?*** 88 earth days. ***Which is longer, a day or a year?*** A year is longer on Mercury than a day. ***Does Mercury orbit in a circle or in an oval around the sun?*** An oval. ***What is the shape of Mercury's orbit called?*** Elliptical. ***Is it hot or cold on Mercury?*** Hot during the day and cold during the night. ***Why is it so cold at night?*** The side of Mercury that is facing away from the sun doesn't have an atmosphere to retain the sun's heat. ***What kind of planet is Mercury, terrestrial or gaseous?*** Terrestrial. ***What does the surface of Mercury look like?*** Mercury has lots of craters. ***What are some reasons it might look like this?*** Asteroids from outer space may have hit it. ***What would the sky look like if you were on Mercury?*** Black. ***Why?*** An atmosphere makes the sky have color. Mercury has no atmosphere. ***When is the best time to see Mercury?*** Early in the morning or early at night. ***Why?*** It is close to the sun, so it tends to rise and set with the sun.

Lesson 4

Why did astronomers think Venus was a twin of the earth? It is close to the earth and is about the same size. ***What would it feel like on Venus?*** Burning hot. ***What is the atmosphere like on Venus?*** It is thick with lots of poisonous clouds. ***What is special about the rotation of Venus?*** It is opposite that of most planets. ***Have very many spacecraft visited Venus?*** Yes, 22. ***Since we can't see through the thick clouds over Venus, how do we know what the planet's surface looks like?*** Scientists use radar, which can get through the clouds. ***Why does Venus go through phases?*** Because of its orbit around the sun.

Lesson 5

Can you remember the seven things that make the earth able to support life? Its distance from the sun, its size, its rotation, its atmosphere, its tilt, its land, and its magnetosphere. ***Try to explain why those things help us to live on the earth.*** If we were closer to the sun the water would dry up, and we would burn up. If we were further from the sun, we would freeze. If we rotated faster we would have winds too strong to survive. If we were not tilted, we would not have seasons and could not grow crops in the summer in colder regions. Our magnetosphere keeps solar winds from burning us with harmful radiation. ***Why do we have different seasons?*** The tilt of the earth causes us to have more or less direct sunlight, depending on where we are in our orbit around the sun. ***What are the four major sections of the earth?*** Crust, mantle, outer core, inner core.

Lesson 6

What is the atmosphere like on the moon? There is no atmosphere. ***What is the color of the moon's sky during the moon's daytime?*** Black. ***Can you explain why the moon has phases?*** As it circles the earth, we are able to see different parts of the daytime side of the moon. ***What is a lunar eclipse?*** When the earth gets right in between the sun and the moon, casting its shadow on the moon. ***Why are the astronaut's footprints probably still on the moon?*** Because there is no wind or rain to wipe them away. ***How does the moon affect the ocean?*** It pulls on the ocean, causing the ocean's tides. ***How are the tides helpful to the earth?*** They cleanse the shore and keep the ocean from being stagnant.

Lesson 7

What makes Mars look red? Rusted iron in the dirt. ***What is the atmosphere like on Mars?*** Cold and thin. ***What is the surface like on Mars?*** It is red and has craters, volcanoes, and rocks. ***What is the name of the biggest volcano in our solar system?*** Olympus Mons. ***What do you remember about the moons of Mars?*** They look like rocks, and they orbit closer and closer to Mars every day. They are named Phobos and Deimos. ***Can you remember how long it takes Mars to revolve and rotate?*** 687 earth days to revolve, 25 earth hours to rotate. ***What is the weather like on Mars?*** As cold as Antarctica. ***Why do some astronomers think Mars would be a good place to visit and, perhaps, live?*** Scientists want to build a community there because Mars is the most similar planet to the earth.

Lesson 8

What is another name for a comet? Dirty snowball. ***What does a comet leave behind it as it orbits the sun?*** Dust and dirt. ***What happens when a comet's dust particles enter our atmosphere?*** It lights on fire and looks like a shooting star. ***What do people call meteors?*** Shooting stars and falling stars. ***What is a meteor called when it hits the earth?*** Meteorite. ***Where have many meteorites been found?*** Antarctica. ***From which planet did some of the meteorites come?*** Mars. ***Where is the asteroid belt located?*** Between Mars and Jupiter. ***What is the Exploded Planet Hypothesis?*** All the asteroids in the Asteroid Belt were once a planet that exploded. ***Can you give some reasons that this might be a correct hypothesis?*** Craters are found mostly on one side of many planets and moons; comets look like asteroids and could be pieces of this planet; Mars has the deepest craters and is the closest planet to the asteroid belt; many moons look like comets and asteroids.

Lesson 9

Do you remember what chemical we need in the atmosphere to breathe? Oxygen. ***How old would you be on Jupiter if you were 48 earth years old?*** 4. ***How does Jupiter protect our planet?*** Jupiter's gravity attracts comets which might otherwise hit the earth. ***Why is Jupiter a little like the sun?*** Jupiter produces its own heat like the sun, has a host of satellites, and is mostly hydrogen and helium. ***What is the Great Red Spot on Jupiter?*** It is a giant storm. ***What do you remember about that spot?*** The storm has been raging for three hundred years and is bigger than the whole earth. ***Why does Jupiter have stripes?*** Jupiter's stripes are bands of clouds. ***Name Jupiter's largest moons.*** Ganymede, Europa, Callisto, and Io. ***Why are they called Galilean moons?*** Galileo discovered them. ***Can you describe Amalthea?*** Amalthea is a big jumble of rocks held together by gravity. ***What do you remember about the spacecraft Galileo?*** Galileo explored Jupiter, sending back information. It took 6 years to get to Jupiter and took a roundabout route. It also had a problem with its antenna.

Lesson 10

Are your bathtub toys more or less dense than water? Since they float, they are less dense than water. ***What is Saturn made of?*** Mostly hydrogen and helium. ***Why would Saturn be an unpleasant place to visit?*** Saturn is freezing, has no solid surface, and has terrible storms and hurricane winds moving across the planet. ***Which planet is considered Saturn's twin?*** Jupiter. ***What are Saturn's rings made of?*** They are made of many rocks, boulders, and a few shepherd moons. ***What do shepherd moons do?*** They keep the rings in place. ***How many years does it take Saturn to orbit the sun?*** It takes Saturn 30 years to orbit the sun. ***Why does Saturn look as if it is being squeezed?*** It is spinning so fast that the planet flattens at the poles and bulges at the center. ***What is the name of the space mission that is going to Saturn?*** Cassini is the mission going to Saturn to explore its moons.

Lesson 11

What chemical makes Uranus blue-green in appearance? Methane. ***Why does Uranus look like a ball rolling around the sun?*** Because it is turned over on its side. ***What makes it look like a loose wagon wheel?*** Its rings wrap around it vertically. ***Why was it so exciting to discover Uranus?*** A planet had not been discovered since ancient times. ***Who discovered Uranus?*** William and Caroline Herschel. ***How were they educated?*** They were homeschooled. ***How long does it take Uranus to orbit the sun?*** 84 years.

Why was Neptune discovered? Astronomers were looking for it. ***What made astronomers think there was another planet beyond Neptune?*** Neptune's orbit wobbled as if it was being pulled by another larger planet. We now know that there is no planet. ***What chemical gives Neptune its blue color?*** Methane. ***Is Neptune the 8th planet from the sun? Explain your answer.*** Sometimes Pluto is the 8th and Neptune is the 9th, because Pluto's orbit crosses over Neptune's orbit. ***How long does it take Neptune to revolve around the sun?*** 163 years. ***What was the Great Dark Spot?*** A storm on Neptune. ***What is the name of Neptune's biggest moon?*** Triton. ***What are geysers?*** They are holes in the ground through which chemicals underground spew. ***Is water coming from the geysers on Triton?*** Probably not.

Lesson 12

What is the Kuiper Belt? A belt of comets orbiting the sun near Neptune. ***How was Pluto discovered?*** Astronomers were looking for a planet that was making Neptune's orbit wobble. ***What are some of the strange features of Pluto?*** Large moon, elliptical orbit, crosses over Neptune to become the 8th planet for 20 years, orbits on a tilt rather than a plane, looks like a comet. ***Why do some astronomers not believe Pluto is a planet?*** The strange features listed above make it different from the other planets. ***What do they think it is?*** A comet or a Kuiper belt object.

Lesson 13

Our sun is a G-4 V star. Can you tell me exactly what that means? It is a medium-hot star (that's the G), a medium-bright star (that's the 4), and a main-sequence star (that's the V). ***Why do you see different stars during different times of the year?*** The earth revolves around the sun and the night sky faces different sections of the universe in its orbit. ***Which group of stars is always present in the night sky of the Northern Hemisphere?*** The Little Dipper. ***What is the name of the North Star?*** Polaris. ***What is special about the star named Sirius?*** It is the brightest star. ***What is a black hole?*** A collapsed star that is so small and massive that its gravity pulls in everything (even light) near it. ***What is a supernova?*** An exploding star. ***What are the three star categories?*** Temperature, brightness, and total energy output (size). ***What is a galaxy?*** A large group of stars. ***In which galaxy is the earth?*** The Milky Way. ***What is the shape of our galaxy?*** Spiral. ***What is a constellation?*** A pattern of stars in the sky that make a dot-to-dot picture. ***How are constellations used today?*** They help us identify where stars, planets, and other celestial objects are located. ***What is the difference between astronomy and astrology?*** Astronomy is a science, while astrology is a false religion.

Lesson 14

What was the name of the first artificial satellite? Sputnik. ***What was the race called between Russia and the United States?*** The Space Race. ***Why was the U.S. worried about Russia's space program?*** It was afraid that Russia could take over the world if Russia had more advanced technology. ***What did the U.S. do first?*** Put a man on the moon. ***What did Neil Armstrong say when he stepped on the moon?*** "That's one small step for man, one giant leap for mankind." ***What is a space station?*** An artificial satellite where people live and work. ***What is the name of the best space station?*** The International Space Station. ***What is the job of people who live on this space station?*** They are scientists doing experiments. ***What is life like on the space station?*** Astronauts float around with no gravity. They sleep tied to the wall, eat out of pouches, take showers with a squirt gun, and drink out of a bag or tube. They must have air, food, air conditioning, and heat. ***How do you become an astronaut?*** Take a lot of math and science classes. Specialize in your favorite area of science, because NASA needs scientists in every field.

INDEX